The economics of irrigation

IAN CARRUTHERS

Reader in Agrarian Development
Wye College, University of London

AND

COLIN CLARK

Former Director
Agricultural Economics Research Institute
University of Oxford

With a Foreword by G. H. PETERS
Director, Agricultural Economics Research Institute
University of Oxford

THE ENGLISH LANGUAGE BOOK SOCIETY
and
LIVERPOOL UNIVERSITY PRESS

First published 1981 by

LIVERPOOL UNIVERSITY PRESS
Press Building, Grove Street, Liverpool L7 7AF

ELBS edition first published 1983

British Library Cataloguing in Publication Data

Clark, Colin, *1905*
The economics of irrigation. – 3rd ed., completely revised and reset.
1. Irrigation – Economic aspects
I. Title II. Carruthers, I. D.
338.1′62
ISBN 0-85323-464-7
ISBN 0-85323-474-4 ELBS

Pbk

Text set in 11/12 pt VIP Times, printed and bound in Great Britain at The Pitman Press, Bath

FOREWORD

This book represents an extensive revision and updating of Colin Clark's *The Economics of Irrigation,* the second edition of which was issued by Pergamon Press in 1970, and which has been out of print since 1973. Dr Clark can justly be described as one of the great pioneers of empirical economics. His earliest book, *The National Income 1924–31,* appeared when he was University Lecturer in Statistics at Cambridge. There had been some previous work on national accounting, but it is no exaggeration to say that Dr Clark, in that book, and in *National Income and Outlay* (1937) placed the subject on firm conceptual as well as empirical foundations. It was all the more remarkable for being done virtually single-handed, without the help of modern computational aids.

Subsequently, with the publication of *The Conditions of Economic Progress* in 1940, Dr Clark was the first to begin international comparisons of living standards and productivity growth. It was in that book, which went through three editions, that Dr Clark's unique style of working became most evident. This can best be described as involving the collection of material for all available countries for all possible time periods (the last edition of *Conditions* contained a telling appendix on the economics of the ancient world), allied to the even more daunting task of distilling a coherent message from what, in less skilled hands, would be a mere jumble of unrelated facts. The technique has remained as the hallmark of a long and distinguished career which took Dr Clark to Australia until his return to the Directorship of the Institute of Agricultural Economics at Oxford in 1953. He held this post until his retirement in 1969, when he again returned to Australia to hold research appointments, first at Monash University, and later at the University of Queensland.

Colin Clark has never ceased to be a pioneer. At Oxford he produced his major study in demography, *Population Growth and Land Use,* and extended his efforts to less developed countries in *The Economics of Subsistence Agriculture,* written with Margaret Haswell. It was also during this period that the first edition of this book appeared. Its focus, again on a world scale, was to study economic returns to the application of irrigation water, with the economics being placed against the background of detailed analysis of the complex scientific factors involved.

This present edition has been compiled with the assistance of Dr Ian Carruthers, Reader in Agrarian Development, Wye College, University of London, who has had extensive field experience in Pakistan, East Africa, and the Arabian Peninsula. The book retains its emphasis on the principles which are relevant in planning irrigation investment and operation, with a particular focus on ways of achieving economic efficiency. The reader will find that this theme is covered in detail in chapters 4 to 9, which deal with groundwater economics, costs of irrigation, economic returns, charges for water and programme planning. This material again rests on the foundation of three chapters devoted to technical aspects of irrigation practice. Such an approach will be of obvious interest to irrigation specialists, and to agricultural and development economists. It should also appeal to economics teachers looking for telling applications of basic micro-economic principles.

My present position makes me acutely aware of the many contributions which Colin Clark has made to economics and agricultural economics and it is a privilege to be asked to introduce work inspired by him. At a more personal level, in an eight year spell at the Oxford Institute beginning in 1959, I was fortunate enough to act for much of the time as his research assistant in a wide variety of projects. It was during this period that Dr Carruthers became one of the many able students attracted to Oxford by Dr Clark's presence. There are many, like me, who have savoured their contact with his vigorous mind and who would wish to acknowledge a debt of gratitude to him.

G. H. PETERS
Director
Institute of Agricultural Economics
University of Oxford

ACKNOWLEDGEMENTS

In preparing this book, we have benefited from contacts with, and comments from, a wide range of specialists. In the field of agricultural economics we have profited particularly from discussions with Eric Clayton, Geoff Allanson, Richard Palmer–Jones, Paul Webster, and other colleagues at Wye College. Jim Dempster, Chris Finney, Don Milne, Roy Stoner, and Andrew Weir have provided insights from their wide and varied experience in the agricultural and engineering consulting profession.

In the area of public health, we have gained from work in the rural drinking-water supply field, which faces many analogous problems to irrigation, carried out with David Bradley and Richard Feachem at the Ross Institute of Tropical Hygiene in London.

From the International Organizations, Graham Donaldson, Peter Naylor, and Gunnar Schultzberg, in particular, have added to the range of experience from which we have tried to draw general insights. Anthony Bottrall of the Overseas Development Institute, and Robert Chambers and Robert Wade of the Institute of Development Studies at the University of Sussex, have been stimulating and constructive critics of our work. Carl Widstrand, Director of the Scandinavian Institute of African Studies, injected the proper degree of scepticism at a crucial time.

We are indebted to our correspondents who have recorded, often in considerable detail, numerous irrigation events and performances, and to the students at Wye College from many parts of the world who have added to our experience and demonstrated the need for a text of this kind.

We would also like to recognize the generous editorial and typographical assistance given by Mr J. G. O'Kane, Secretary of Liverpool University Press and Mr Bernard Crossland, typographical advisor to the press, who saw this book through all its stages. A special vote of thanks goes to George Peters, who reviewed early drafts of the book in a very thorough, incisive and helpful manner, and provided the Foreword to the book.

IAN CARRUTHERS
Wye College
University of London

March 1980

CONTENTS

LIST OF TABLES

LIST OF FIGURES

CHAPTER 1 THE ECONOMICS OF IRRIGATED AGRICULTURE

Irrigation projects generally endear themselves to agriculturists because they tend to promote maximum yield per hectare—a well-understood and indeed, cherished goal. They also reduce risk assignable to weather. Irrigation projects have, in addition, many attributes that satisfy the objectives of politicians; particularly a rapid, visible, and dramatic impact, and the tendency to be closely associated with a political promoter. Donors of economic aid favour irrigation projects for similar reasons. Engineers enjoy the challenge of designing irrigation schemes, particularly when they are on a large scale, and therefore speak of water 'wasted' when it runs into the sea; if it runs into the sea through a good dam site or a desert they become almost uncontrollable. Development administrators see irrigation as creating opportunities for enforcing discipline in production, marketing, and finance. This somewhat cynical introduction is presented in order to indicate the powerful backing which irrigation proposals may generate independently of their economic merit.

ECONOMIC CONCEPTS IN IRRIGATION PRACTICE

Typically irrigation projects consume large quantities of extremely scarce resources, such as capital and recurrent finance as well as skilled man-power. Therefore the concepts and methods of economic science are very relevant in weighing alternative forms of irrigation projects and alternative designs, timing, location, and operating procedures. Economic considerations are especially timely today because the new agricultural technology that stimulated the Green Revolution requires optimal soil moisture for effective utilization. In the rush to join in the transformation of agriculture, there is a danger that countries may overlook or relegate economic considerations and thus the farmers' promised banquet may turn out to be simply an engineering picnic.

Notes and references for this chapter begin on p. 248.

Furthermore, the emergence of widespread unemployment and the continuance, and even the growth in some countries, of abject rural poverty, has resulted in a rash of irrigation schemes with the main objective to settle landless farmers. Decisions are often taken largely on the basis of a vision of what could be achieved by irrigation. In this approach the nature and magnitude of the unemployment problem is all-important, and the merit of the means selected to solve the problem escapes careful economic scrutiny. The powerful momentum built-up in the past behind largely irreversible, and expensive irrigation proposals, provides a warning and a justification for more economic analysis in future.

The era of the big dam is, with one or two notable exceptions, now over. But many potential irrigation developments remain. In Asia the World Bank notes:

> About a quarter of India's present farmland is irrigated and it is known to be possible to raise this to a half. The development of Pakistan's irrigation potential is much further advanced after a century of irrigation from the waters of the Indus river system, but more land could be irrigated, especially with further groundwater development. In Bangladesh, water resources are abundant but unruly: long, dry periods are followed by flooding from the enormous Ganges-Brahmaputra-Meghna river systems which spread out over the vast, flat alluvial deltas. Drainage, flood control, and minor irrigation projects are required to harness the water resources more effectively. In Indonesia there is potential for doubling the present irrigated area with the construction of new gravity systems, and development of swamps and tidal land outside Java. Extensive groundwater systems can also be exploited in the longer run. Burma has a large potential for rapid expansion of high yielding agriculture, especially in the reclamation and development of the vast paddy regions of lower Burma, where there are still about a quarter of a million hectares of abandoned ricelands.[1]

This book is addressed to all those who may have any responsibility for spending money on irrigation, whether for small schemes or large, whether private or public, whether in arid or in humid climates, and whether their responsibility is direct, or the indirect responsibility which falls upon those who help to form political and business opinion. It will be of interest to the irrigation agronomist, the engineer, the administrator, and the planner. It will provide useful field insight for the economics student who all too often is struggling to find the practical relevance of the obscure, unrealistic principles and models which form the core of much modern economics teaching.

It does not deal at length with the theory or project appraisal or water charges, or of the allocation of water to different uses, subjects on which there are many writings of great refinement. Unfortunately refinement and relevance are not inevitably linked. This book is intended as a piece of applied economics, setting out the basic concepts and principles, and collecting the critically important facts over as wide a range as the authors have found possible.

In this text, knowledge is presupposed of only the simplest principles of economics, among them of the distinction between money and real costs, of the distinction between marginal and average costs and products, and of the applicability of the principle of opportunity costs.

Collation and appraisal of irrigation experience are made difficult by the paucity of published and available literature on specific schemes, despite the vast sums invested in capital and operation of irrigation projects. There is very little public scrutiny of, or debate on, proposals for irrigation projects. In normal planning practice documents are prepared by consultants or government departments for the government client. Although such documents are often extremely comprehensive—for example, the Lower Indus Project ran to twenty-eight volumes—criticism is seldom invited and often, indeed, precluded by various security curtains. This attitude also pervades the operation phase. Past investments are seldom evaluated, and therefore planning groups often lack feedback of vital experience when formulating new projects.[2]

ERRORS OF ANALYSIS IN ASSESSMENT

Economic analysis is sometimes considered to be an unnecessary prerequisite for irrigation development. A recent paper to the Tenth Congress of the International Commission for Irrigation and Drainage noted: ‘Irrigation systems . . . in arid regions, where farming is impossible without irrigation . . . seldom requires any particular economic examination before design’.[3]

It is not only in centrally planned economies that such views are taken. Casual analysis or unjustified claims are often made for particular irrigation projects, for a variety of political or economic reasons:

> Because water supply decisions are largely made in the political arena rather than in the market place, there is a great incentive for special-interest groups to obscure the real issues involved in government-subsidised water projects and to exploit public romanticism for 'making the desert bloom even as the rose', thus obtaining public support for their own financial gain.[4]

Hirshleifer and Milliman[5] are more specific:

> Rational re-allocation of existing supplies of water, such as would occur if prices were raised in response to scarcity, is almost never even considered as an alternative to new construction.

Such reasoning for unjustifiable claims can be summarized as follows:

(1) Politically determined prices are inflexible. New York, for example, at the time when Hirshleifer and Milliman wrote in 1967, had not modified their price schedule since 1933.

(2) Political rigidity makes it commonly impossible for one water jurisdiction to sell title or rights to another. Thus there exists an incentive to rush into construction and to nail down and exploit available supplies.

(3) Certain errors in economic reasoning have played a role: among them, ignorance of the marginal principle, double counting of benefits, and the use of inappropriately low discount rates. Some advocates of irrigation not only fail to understand economics but also the simplest principles of accounting, and are unable even to distinguish gross from net returns.

However, these analytical errors have had much less practical significance than what might be called the non-analytical error. This is the belief, usually quite unconscious, that there are fixed 'needs' or 'requirements' for water rather than economic demands. No matter how conclusively it is refuted by facts—for example, the experience, which is all too common, of the inability to sell high-priced water—the belief that demands are absolutely inelastic continues to dominate much planning in this field.

Little attention need now be paid to the secondary economic effects of large irrigation schemes, which have been supposed to be unique and beneficial. There was some justification for making such a case in the industrialized countries in the 1930s, against a background of severe unemployment of labour and other productive resources, which may have been mitigated by the indirect or multiplier effects stemming from a large irrigation scheme. It is now much more clearly appreciated that it is very seldom that governments are in a position where general excess capacity exists within the economy that can only be used by stimulating demand by irrigation investment. All projects produce indirect effects and it is necessary to consider the *relative* merits of irrigation, and of alternative projects for spending the same amount of capital or other resources, which would also give much the same beneficial secondary effects. In much of the world, except for occasional periods such as those following the 1973 oil crisis, most productive resources which are employable are already fully employed, or over-employed, and the secondary economic effects of large capital expenditures may even be harmful. The economic impact of irrigation schemes must now be judged, therefore, largely on their direct economic costs and returns.

THE PLANNING PROBLEM

Irrigation may be defined as the application of water, by human agency, to assist the growth of crops and grass. In general, the word does not cover the other purposes for which man stores and controls the flow of water—that is to say, for household and industrial water supplied, for hydro-electric power, for the provision of navigable channels, and also the elimination of unwanted water, in flood measures, and in the draining of swamps and waterlogged land. These other purposes, however, may sometimes share with irrigation the use of works, equipment, waterflows, and administrative organization; and they therefore have an indirect effect upon the economics of irrigation. Water for irrigation may be, firstly, pumped from underground sources by means of wells; secondly, drawn from the natural flow of streams; and thirdly, obtained by damming or otherwise regulating the flow of streams. It may be applied to the crop by flooding, by channels, by spray, or by drips from nozzles.

Irrigation is generally advocated because deficiencies of climate constrain agriculture. Total rainfall may be inadequate or unreliable. Socio-economic reasons may also be utilized by advocates of irrigation, such as meeting regional development or income re-distribution goals, 'settling nomads', relieving population pressures, creating employment opportunities, or reducing or avoiding famine risks. Technical-economic reasons may be employed to justify irrigation, such as the fact that the degree of control over plant growth increases crop quantity and quality, and facilitates low-cost processing. Irrigation may also enable governments to exercise greater control over farmers' cropping decisions through extension services and possibly make it easier to regulate agricultural taxation. Given such a wide range of possible goals it is important that objectives should be clearly indicated before design and appraisal are undertaken.

ESTIMATION OF SYSTEM COMPONENTS

An irrigation project may have all or some of the components shown in fig. 1.1. This is derived from a study in Bangladesh,[6] but it is equally applicable to large parts of India and Pakistan, and it has some general relevance. The components are crops and their requirements for evapotranspiration. Water sources are either river, reservoir, precipitation, pumped fresh groundwater, or pumped and diluted saline groundwater. Loss of water occurs not only beneficially to crops but also, wastefully, to seepage from rivers, canals, field water courses, and fields. This seepage, however, recharges the aquifer, whence it may be recovered by tubewell pumping. Some groundwater may return to the river in periods of low river levels, if the water-table level is high. Water is also removed by gravity or pumping after it has run off from the surface of fields and non-irrigated areas. This outline of the major flows in the system is completed with mention of evaporation from rivers, canals, and the soil, and evapotranspiration by crops and weeds. Discussion in this book covers measures appropriate to these variables in this system for irrigation projects on both arid and humid land; it also considers relevant factors in estimating economic and financial costs, and benefits for developing these systems in countries with a wide range of environments.

FACTORS IN IRRIGATION PLANNING

To the farmer water is simply one input in the production process. However, this input is rarely available in the ever-changing quantities required to meet agricultural demands. There are generally several feasible technical solutions to the problem of matching seasonal water demands with available and potential supplies. Irrigation planning involves the identification of alternative means of artificially achieving a satisfactory moisture regime and the allocation of scarce investment resources among these alternatives. Therefore, whatever the major goal, each proposed

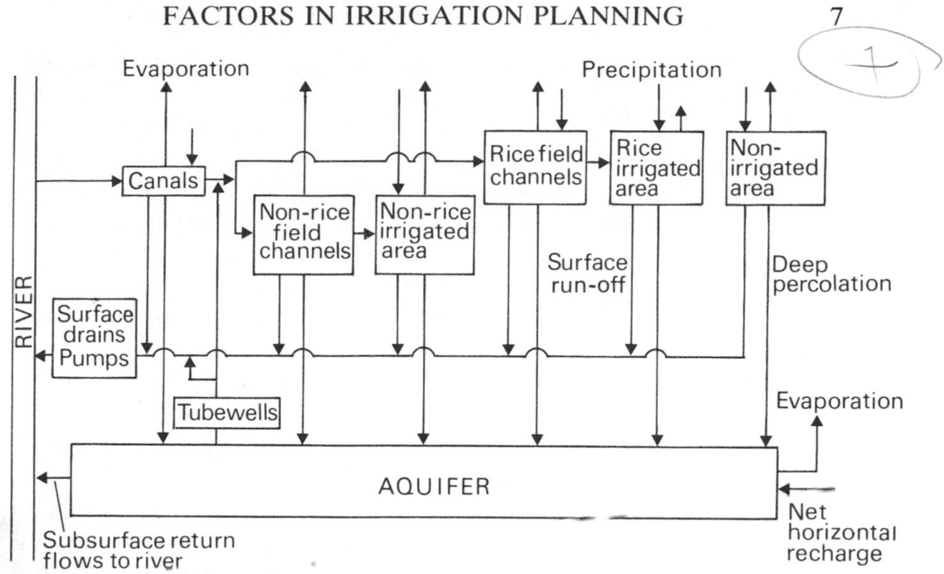

Figure 1.1. Flow chart of water balance for an irrigation project.

water development has to be subjected to technical and economic tests to prove its efficiency. On the technical side this involves consideration of such issues as plant–soil–water relationships, water sources and means for their development. Economic tests or appraisal necessitate consideration of such issues as plant-soil water relationships, water sources and means for their development. Economic tests or appraisal necessitate consideration of spatial and temporal allocation of investment resources, which involves decisions on discount rates and forecasts of benefits and costs. Increasingly, as equity goals are emphasized in public policy, the incidence of benefits must be assessed.

Some of these issues are considered below in elementary and general terms. Among those indicated are, firstly, the nature of plant response to water, including the quantity and quality limits; secondly, alternative sources of water; thirdly, some important technical and economic features of water development; and, lastly, the relationship of agricultural institutions and development.

Plant response to water

Crop growth is dependent upon water which is generally derived by abstraction from soil moisture reserves.[7] These reserves may be built up by rainfall or irrigation. To achieve maximum plant growth, soil moisture levels should fluctuate only within a fairly narrow range between field capacity and permanent wilting point. Field capacity may be loosely defined as the moisture status of a free draining soil when it is holding all the water it can against the force of gravity. However, soil drainage is a continuous process, and there is no point at which a unique quantity of water is held against gravity. However, for convenience, field capacity is often defined as the quantity of water held at a particular suction pressure and it is measured at a specific time after irrigation, generally forty-eight hours. There has been dispute among plant physiologists as to whether soil water is equally available over this whole range or whether growth is affected by the higher soil moisture tension as permanent wilting point is approached. It is now generally held that crop growth is impeded even with a temporary high soil moisture tension.

An ideal pattern of water availability is seldom achieved in practice, least of all under rain-fed conditions. Even for irrigated agriculture, it is difficult to match supplies exactly to seasonal demands of crops. Requirements of water for crops generally rise to a peak as the crop reaches maturity, but in many cases it is not possible to meet this variable demand because irrigation facilities are limited by technical constraints to a specific discharge. The problem in management for a farmer is to obtain a profitable combination of crops and sowing dates that match a given discharge. Although it is theoretically feasible to find such a cropping pattern, it is unusual to do so, and there are generally either unsatisfied peak demands or excess irrigation capacity for a part of the year, or both.

An irrigation regime that provides soil moisture for maximum crop growth and yield per hectare will be unlikely to produce maximum output per unit of water. It will be shown in chapter 2 that the resource availability and the shape of the crop response curve under peasant farm situations is such that, in most instances, a farmer will be acting rationally to under-water crops and accept a lower yield per hectare. He will then achieve higher total output from an increased area cropped, assuming that irrigable land is not a constraint.

WATERLOGGING

Another biological feature which has important economic implications is that there are limits to the time for which crops (other than rice) can tolerate a saturated soil. Ideally an irrigated area should be protected against periodic flood damage, but the costs of such protection are generally high. A more wide-spread problem than occasional flood damage is waterlogging of the soil, which impedes most crop root development. Waterlogging may be caused by a number of factors, including flooding. Perhaps the most serious waterlogging problems occur where there has been a general regional rise in the level of the groundwater table. This is often a consequence of increased seepage following the development of large-scale irrigation. In the Sukkur Barrage Command in Pakistan in April, 1932, only 15 per cent of the land had a water-table above 3·7 metres, but by April, 1964, more than 65 per cent of the area was in this condition and this general rise was continuing.[8]

QUALITY OF IRRIGATION WATER; SALINITY

In arid zones growth of crop under irrigated conditions is subject not only to consideration of timing and quantity of supplies, but also to quality factors. Irrigation water invariably contains small quantities of dissolved salts, some of which are plant nutrients. These salts should, however, be present only in low concentrations, and in practice water with salt content of more than two to three thousand parts per million is likely to be toxic to most plants. The salt concentration in groundwater often exceeds these limits, and it should be mixed with fresh river water before it can be used for irrigation. Occasionally groundwater with total dissolved salts below maximum concentrations proves valueless because of harmful amounts of a single ion such as sodium or boron. Excess sodium can damage soils, and excess boron is toxic to all plant life.

If the total rainfall plus irrigation water are insufficient to leach residual salts from the crop root zone, then salt concentrations will in time build up to harmful limits however small the quantities of salt that are brought in at each irrigation. Salts are also often deposited following evaporation of slightly saline water from the capillary fringe of a high water-table. In planning irrigation development for arid areas, provision has to be made for leaching of existing and future deposits of salt, and adequate control of the groundwater table must be arranged to restrict upward movement of salts.

Sources of water for crop production

Water is supplied to the soil from one of three sources: rainfall, surface irrigation supplies, and groundwater. These supplies are normally intermittent, but the crop demand is continuous (with diurnal fluctuation). The soil acts as a reservoir and buffer against dehydration.

It is rare for the seasonal pattern of precipitation to coincide with crop water requirements, and often, even in humid areas, short-term deficiencies in soil moisture can occur which check crop growth and reduce financial returns. In many parts of the tropics and in monsoon areas of Asia extreme seasonal differences in rainfall occur. For instance, in Bangladesh 85 per cent of the rainfall occurs in summer, between May and September, and in these circumstances winter irrigation can be profitable. A study in the tea area of Sylhet District, where annual rainfall exceeds 3300 mm, has shown irrigation in the dry season to yield high rates of return.[9] In Britain, although evaporation is generally less than precipitation, research at Rothamsted has shown a need for irrigation five years out of ten throughout southern England.[10]

If rainfall is not adequate or dependable, the technical and economic feasibility of supplementing these supplies has to be considered. The first surface water developments are usually simple diversion structures in the upper reaches of river valleys, but these cheap sites are soon exhausted. Downstream diversion structures or barrages often enable large areas of river basins to be commanded by irrigation canals. As an alternative to a barrage it may be possible to lift water from rivers by low-lift pumps at a lower initial capital cost but with much higher operating cost.

The basic sources of river water are rain and snow, but because of uneven rainfall patterns and rapid snow melt it is exceptional for the resulting river discharges to fit the requirements of agriculture. It is, of course, technically possible to store river water at times of surplus for release during periods of shortage. However, surface reservoirs for irrigation water storage are generally expensive, subject to seepage and evaporation losses, and, when directly on a river, liable to rapid and permanent loss of capacity due to the deposition of sediment.

The value of an irrigation scheme should not be confused with its scale. In mountainous arid areas or in savannah pastoral areas small-scale projects can be a locally very important source of food, fodder and employment. Scholz[11] describes changes in the way of life of nomadic herdsmen and irrigation farmers in

Baluchistan and the shifts between these occupations following disaster (e.g. loss of a spring after an earthquake). Fresson[12] describes a survey of small-scale paddy irrigation schemes in Matam, Senegal which were completed with a high degree of public participation and self-help. Unlike many self-help schemes they also appear to be self-sustaining and profitable. The costs are one third to one seventh major irrigation projects and paddy yields average 3·4–4·2 ton/ha.

Groundwater may be exploited by means of shallow, hand-dug wells, or deep tubewells. It is a useful source of irrigation water where there are good sub-soil aquifers (for example, sand or gravel), the water-table is not too deep, and the groundwater is fresh. Groundwater storage is not subject to significant evaporation losses, and because the water is used close to the well, conveyance losses are reduced to a minimum. Wells have flexibility in that they can be operated at any time according to farm demand. An important feature of a well, where the aquifer is not confined by an impervious layer above the water-table, is that its operation lowers the water-table and drains the soil. Thus the pumping of tubewells in excess of the groundwater recharge will lower the groundwater table. If this process is continued then the stored water in the aquifer is 'mined'. As pumping costs are proportional to the height to which water is lifted, a point must eventually be reached where pumping costs exceed the agricultural value of the water. Groundwater economics are the content of chapter 4

Technical and economic aspects of large-scale irrigation

Several of the technical issues considered in water resource planning have important economic implications. As an illustration, four decision areas are discussed briefly here in general terms:

(1) where and when to release irrigation supplies;
(2) how to cope with peak demands;
(3) how to integrate individual project areas to achieve maximum benefit;
(4) how to benefit from the new agricultural technology, which is the basis for the Green Revolution.

It is possible to spread irrigation water over a relatively large geographic area and achieve low cropping intensities but at the same time benefit a large number of farmers. If the water is spread thinly over spare land, this may cause salinity because there will be insufficient leaching. Alternatively, the same resources may be concentrated on a small, intensive area; this is much cheaper as there are economies in water use, canal construction, and services. However, there are a relatively small number of beneficiaries, which may render this course less acceptable politically. If water is concentrated on a small area, drainage will be required or the water-table may rise, leading to waterlogging or salinity or both. Decisions often have to be taken on the timing of supplies, such as, in the short-term, when to release stored water from reservoirs and, in the longer-term, whether or not to 'mine' groundwater.

River discharges seldom correspond to an ideal pattern over the year, for they are generally peaked; but even in the rare situations where river flows match crop requirements, there is likely to be a constraint on irrigation capacity. Unlined canals work efficiently only at or close to a specific discharge. In canal design, maximum flow should be only 1·6 times the minimum. A dual canal has recently been designed for the Lower Indus which can achieve an overall range of 3:1 maximum:minimum flow, but this is exceptional.[13] Pumps lifting river water or in tubewells also have a design discharge. Thus peak requirements can be met only by additional canals or pumps to be utilized for short periods of the year. In the case of canals there are construction and operation problems if additional channels are built to supplement supplies to an area with an existing network of irrigation and drainage channels. The economic problem is to compare the estimated cost of the additional capacity with the forecast benefits from an increased area cropped and a more profitable cropping pattern. Alternatives which have to be considered are as follows:

(1) can existing supplies be more economically used at crucial times?

(2) can crops draw upon soil moisture that was applied before sowing and stored in the soil profile?

(3) can moisture stress and subsequent yield reduction be accepted by farmers?

(4) would it be more appropriate to modify cropping patterns and sowing dates to achieve a better match between water supplies and requirements?

It is characteristic of large-scale irrigation developments that because there is a great measure of physical interdependence between different users of water resources, these resources must be planned and operated as part of an integrated system, in order to achieve maximum benefit. Conflicts of interest may occur among users. For instance, in a hydro-electric scheme, scarce water supplies may be required at different times of the year both for power generation and for agriculture. Control of rivers for irrigation may harm fishing interests. In the Lower Indus, fishing has suffered because there is no flow in winter, all water being abstracted for irrigation. Barrages have also hampered the summer migration up-stream of an economically important fish, hilsa. In the Nile basin the disruption of fisheries has been more serious as a consequence of the construction of the Aswan High Dam. Sardine fisheries in the marine delta have been all but eliminated because of the loss of Nile nutrients. This loss of 18,000 metric tons per year has not been offset by increased freshwater fish yields from Lake Nasser above the dam[14] Indeed, the experience in Kariba and elsewhere suggests that, for reasons which are not fully understood, the freshwater fish yields are likely to remain much below preconstruction forecasts.

The conflict of interest may extend to the general public who are not direct water users. The increased health risks associated with irrigation are well known. They are poetically summed up by Kenneth Boulding in this verse:

> Then the water in the lake, and what the lake releases,
> Is crawling with infected snails and water-borne diseases.
> There's a hideous locust breeding ground when water level's
> low,
> And a million ecological facts we really do not know.[15]

In the main, however, there are substantial positive benefits to be achieved by over-all development of water resources. These benefits are derived from investment in complementary facilities for irrigation, flood protection, power generation, water supply, and other uses. The objective of the planning procedure is to develop methods to appraise alternative plans and to achieve optimum sequence of investments. It is partly because the benefits to be derived from planning water resource projects on a large scale and as an integrated sequence are potentially high that water resource development is in most developing countries considered to be an area of government responsibility.

Each proposed water resource development requires rigorous technical and economic appraisal. This may appear obvious, but often, in arid areas, water is considered to be a unique resource, and, therefore, engineering or agronomic feasibility are considered sufficient conditions to justify investment. There are several reasons why this may lead to a misallocation of resources. The various parameters that go to make up cost and benefits are extremely variable through time. For example, irrigation developments become progressively more expensive as less favourable sites are developed. There may also be a higher rate of inflation affecting capital construction costs than there is generally throughout the economy. It might also be considered that recent rapid advances in agricultural technology (for example, plant breeding, fertilizers, and crop protection) may have created attractive alternative areas for agricultural investment.

On the other hand it is well established that these advances, which have been the basis for the Green Revolution, require assured crop water supplies for economic exploitation. Irrigation water complements other modern technical inputs. When water, fertilizers, crop protection, new improved seeds, and so on are simultaneously applied, the yield response is greater than the sum of the response to each individual input if it is applied independently. This has proved to be practically important for wheat and, to a lesser extent, rice in Asia.

Technological advances have also occurred in the engineering aspects of irrigation. For instance, there have been dramatic developments reducing cost in earth moving and tubewell construction technology in the last decade.

It may be concluded that irrigation proposals require careful and continuing economic appraisal with engineering and agronomic feasibility considered necessary but clearly not sufficient conditions for investment. Each proposal has to pass economic tests that indicate from current information that the proposal is economic, namely, that it is at least equally productive as alternative uses of the same resources. Procedures for this type of appraisal are discussed in chapter 8.

AGRICULTURAL INSTITUTIONS AND IRRIGATION DEVELOPMENT[16]

Modern developments in the scientific understanding of the need for, and role of, water in agriculture, of the complementary inputs required to make good use of water, and of skills, know-how and equipment necessary for irrigation, have great potential for making better use of water. Recent experience, however, suggests that without an adequate institutional framework the contribution of technological innovations to either agricultural productivity or the welfare of the rural poor will remain minimal. In too many cases innovations have failed to achieve the cropping intensities, levels of yields, productivity, incomes, and welfare which are reasonably expected of them. Irrigation schemes often fail because they neglect the role and special problems of agriculture in general and small farmers in particular in making productive use of the large-scale engineering works provided publically. Technological developments alone will not be sufficient to promote rural development; they are even unlikely to achieve production goals, except at high cost. Furthermore, although perhaps one-seventh of the world's agricultural land is irrigated or potentially irrigable, the majority of the world's population will continue to depend on rainfed agriculture, and there remains great potential for increasing the productivity of these lands. It is extremely misleading to assert that 'In agriculture, water holds the key to increased production through irrigation.'[17]

Most Third World countries undertaking irrigation projects will depend, however, upon improvement not only on their technological equipment but also on relevant institutions, including those devoted to agricultural research and development. They will need to use to the full the existing international network of agricultural research centres, which often have links with regional research centres and take part in the programmes of international agencies such as UNDP (United Nations Development Programme) and FAO (Food and Agriculture Organization). In addition, there are growing, although still inadequate, links between the international centres and national research, training and production systems.

There is a real danger, well-understood by administrators, that the provision of the high-grade staff and first-class working facilities that are necessary to ensure rapid progress at the international centres will widen the gap between them and local agencies. More importantly, perhaps, this lack of appropriate communications will destroy, at farm level, opportunities to gain knowledge of what development is locally suitable and under what constraints they must expect to operate.

Experience suggests that the dangers of non-replicable enclave research units are more apparent than real, and that in fact agricultural research at all levels represents a very productive use of resources when compared with alternative agricultural investments (Aresvik, 1975,[18] although see Hewitt, 1976[19] for a cautionary tale). The collaboration of social scientists with natural scientists will help to ensure that technology is appropriate to the national resource endowment and compatible with multiple objectives of governments.

It must be emphasized that to be successful, transfer of technology must include not only the hardware but also the software of services, institutions, and attitudes. Some authors contend that the technology adopted largely reflects the interests of those holding power in society (of politicians, bureaucrats, urban consumers, and large landowners) and technology appropriate to the majority may be excluded (Stewart, 1977[20]). Ruttan, 1977[21] contends that technical change in agriculture is endogenously generated but that institutional change is not occurring to keep up with the technical potential. He states that the

> developing world is still trying to cope with the debris of non-viable institutional innovations: with extension services with no capacity to extend knowledge or little knowledge to extend: cooperatives that serve to channel resources to village elites; price stabilization policies that have the effect of amplifying commodity price fluctuations; and rural development programmes that are incapable of expanding the resources available to rural people. . . . Unless social science research can generate new knowledge leading to viable institutional innovation and more effective institutional performance, the potential productivity growth made possible by scientific and technical innovation will be under-utilized.

In a later paper Ruttan shifts his ground slightly and produces a more optimistic hypothesis. He contends that institutions are responsive to economic incentives and therefore suggest that, providing the pay-off or return is sufficiently high, institutions will arise to enable exploitation of the new profitable opportunity to occur.[22]

The present authors concur with this perspective. This review of the economics of water use in agriculture is intended to provide a factual and conceptual basis for a greater understanding of the economic issues by both natural and social scientists jointly engaged in promoting rural welfare by irrigation development.

ALTERNATIVES TO IRRIGATION

Irrigation can, in the words of V. N. Rao, cast a shadow of indifference on the dry land in its vicinity.[23] His surveys showed that in areas with mixed irrigation and dry farming most farmers, particularly larger farmers, positively neglect their dry land to reap the benefits from irrigation. From an economic viewpoint any fall in net value added in the dry areas is a cost to irrigation development.

Although readers of this book will be interested primarily in irrigation the authors wish to close this introductory chapter with a cautionary note. Readers should be aware that alternative agricultural investments also require evaluation. Before expensive irrigation systems are adopted, it should be ensured that rain-fed agriculture cannot be profitably improved.

Varieties of crops can be planted that come to maturity within the rainy season. Crops which grow in the cool season or high-yielding varieties which increase the yield per unit of rainfall should also be investigated. Rainfall conserving or harvesting techniques such as tied ridges can improve the availability of rain-water. Evaporation from soil surfaces is typically 25 to 50 per cent of field losses. These losses can be cut by natural and artificial mulches such as plastic film or asphalt. Evaporation from useless plants such as weeds or neighbouring trees can be reduced to a minimum. Evaporation from crops can also be cut by use of wind-breaks, covering with structures such as plastic or glasshouses, defoliating useless leaves, or applying antitranspirants. These latter materials either close or clog stomata or reflect heat, thereby reducing evapotranspiration needs (see chapter 2). These areas of technology are at present in embryonic stages. Agriculture can often be improved substantially without irrigation.

CHAPTER 2 WATER REQUIREMENTS OF PLANTS

THE ROLE OF WATER IN PLANTS

Where the sun strikes down on the land, it becomes hot. But there are limits to this heating, as to every physical process. As the land becomes hotter, the rate at which it radiates heat back into space increases rapidly (in proportion to the fourth power of absolute temperature); and some of the heat may be carried away by convection (hot winds). These processes, however, sometimes do not attain equilibrium until the temperature has risen to about 50 °C which may be reached at midday in hot deserts. With a few curious exceptions (which will be described below), no plants can live at such temperatures.

However, if water is present, it will evaporate. The evaporation of each gramme of water absorbs substantial quantities of heat. Other climatic factors being given, the presence of water therefore reduces the average temperature considerably, and also reduces the extreme temperature variations found in deserts.

A plant uses limited quantities of water for its own growth metabolism. Nearly all of the water which is transpired by plants is for the purpose of maintaining thermal equilibrium. Unfortunately plant physiologists are still far from being able to agree on the maximum temperatures which plants can tolerate. It is of course clear that some plants are less tolerant of high temperatures than others. It is thought that perhaps plants can, for a few hours at midday, temporarily stand temperatures above their average level of tolerance.

In every plant the leaf surface is covered by minute stomata through which the plant evaporates water. In addition, the stomata allow absorption of carbon dioxide from the atmosphere, providing the basic material for photosynthesis and growth. There is again some disagreement among plant physiologists as to what happens in very hot periods. Some hold that the plant will itself close its stomata at the cost of a temporary rise in leaf

Notes and references for this chapter begin on p. 249.

temperature. However, a more widely held view appears to be that in hot periods of bright sunlight the stomata are open to a 'wasteful' extent, with the plant cooling itself more than is really necessary, at any rate for a short period.

The pineapple is an atypical plant in that it collects a surplus of carbon dioxide through open stomata at night and then closes its stomata in the heat of the day.[1] This reversal of the normal process makes it more efficient in water utilization than other plants. A field of pineapples evaporates more just after planting than it does later with full ground cover.

Cacti and other desert plants are adapted to their surroundings by keeping their stomata almost permanently closed, tolerating high leaf temperatures, growing glossy skins which are impermeable to water and reflect a substantial portion of the incoming solar radiation, and keeping large internal reserves of water. Naturally, their rate of growth is very slow. Plants useful to man have to be grown at much lower average temperatures, and they have to keep their stomata open for most of the day.

Two other functions of water in the plant are the translocation of nutrient materials and the maintenance of turgidity. However, the most vital function for water consumption is the maintenance of a viable thermal regime. Less than 1 per cent of water inputs to the plant are retained in normal growth.

WATER AND CROP GROWTH[2]

Plants and stress

A basic outline of the physiology of crop plants may assist in understanding and guiding policy options in irrigation planning and operation. Generally, plants grow best when they are not under moisture stress. Stress, technically, occurs when there is a loss of water from a plant, or a part of a plant, creating a negative water potential. When the water potential, measured in units of pressure (bars) falls below −0·5 to −1·0 bars, stress is said to exist. However, most physiological processes continue normally until the water potential falls below −10 bars. For most crop plants permanent wilting point is reached by the time −20·0 bars deficit in the soil is reached although there is considerable variation between plant species in the range −5 bars to −200 bars. Plant leaves can generally tolerate higher deficits than −20 bars for short periods (a few hours) but if the deficit is sustained permanent wilting results. Permanent wilting point varies for species being about −19 bars for tomato, −43 bars for cotton and

−110 bars for *Acacia aneura* (Slatyer 1967). Within a species cultivars vary in their drought tolerance. For example, Sullivan and Eastin (1975)[3] showed that of two cultivars of sorghum subjected to stress of −33 bars one exhibited 90 per cent recovery and the other did not recover at all. Some crops appear to benefit from short periods of quite severe stress (for example, coffee just before flowering), but these are exceptions.

Water moves from areas of high potential to those of low potential at a rate determined by the difference in potential and the resistance of the pathway. Parts of plants or plant entities which are at a potential below critical levels are under stress. The effect of water shortage is to promote stress, which reduces crop growth and yield potential. If stress is severe, it may lead to crop failure. As described in the foregoing section, stress develops in plants because water evaporates from leaves, mainly through the stomata, absorbing latent heat, thereby keeping plants cool and therefore photosynthetically active. Evaporation (transpiration) reduces the water potential of the leaves, which consequently draw water from the roots, whose potential also falls depending on the rate of withdrawal of water and the rate at which water can be drawn from the soil. The rate at which water can be drawn from the soil depends on the water potential of the soil in contact with the roots and the extent of root system which can effectively absorb water. As the water potential of plant tissue falls it will tend to wilt, and, as also with soil, the resistance to water movement rises.

The development of stress is affected by stomatal behaviour, which, modified by light intensities, humidity, temperature, carbon dioxide concentration, and endogenously produced plant hormones, is extremely complex. Although, basically, stomata have evolved to minimize transpiration losses consistent with making the best use of light and other environmental variables, it is worth noting that most agricultural plants are grown outside their natural habitat, and it is difficult to predict their stomatal response and to assess whether it is sub-optimal.

Whereas atmospheric conditions—temperature, humidity, and so on—as modified by the extent of aerial parts and stomatal behaviour are the *demand* aspects of the development of stress, the *supply* side is determined by soil moisture conditions and the development of the root system. The higher the soil water potential (see further below) and the more extensive the root system, the greater the rate at which water can be drawn into the plant, and therefore the lower the stress that will develop in the plant for any given rate of demand. But because water moves slowly in soils that have dried somewhat (and also in the plant), it is possible for considerable quantities of water to remain in the soil and yet not be readily available to plants because of the large difference in water potential that must exist to move this water into the plant. Under these circumstances, on hot days with a high rate of evaporation, plants will become severely stressed and growth will be limited despite apparently plentiful supplies of soil moisture; indeed, plants can be heavily stressed even when the soil water is at its maximum potential if the evaporative demand is high enough. Different species and varieties become stressed to different extents, and the effect of given levels of stress varies.

The plant draws water from storage in the soil; it draws it mainly from areas where the roots are densest, and it is these areas which have the potential for supplying water to maintain moderate levels of stress provided demand is not excessive. Water in less densely rooted zones is not readily available to meet high rates of demand; thus, although it can permit plant survival (or overnight recovery), it will not allow much growth if high rates of demand are normal. Therefore, if plant growth is desired, it will be important to maintain the water potential in the densely rooted zones at a high level. This is probably largely responsible for the high productivity of irrigation schedules which apply small quantities at high frequency, and of drip irrigation systems.

Of course, different plant species and varieties are adapted to different conditions and respond differently to different environmental situations. Also, the effects of stress on yield of useful plant products depends on the (physiological) stage of crop growth at which it occurs. For example in maize, stress at tasselling results in a greater reduction in yield than stress at other stages in the crop's growth.[4] Clearly, the effects of stress will also depend on the physical state (health) of the crop at the time when the stress occurs; a healthy plant is far more likely to survive periods of stress, but this does not mean that reducing stress in a healthy plant will have a bigger effect on output of useful products than using the same quantity of water to reduce stress on a less well developed plant (at the same stage of growth), though this has to be investigated empirically. In the absence of the necessary experimentation, judgement, experience, and agronomic expertise as well as an awareness of the economic opportunity costs of water applications will be required.

Soil moisture may be replenished by rainfall or irrigation, which leads to downward percolation as the soil moisture capacity is sequentially replenished from the top;[5] or it can be replenished by upward or sideways movement of water (sub-irrigation).

Plants can draw on stored water; its quantity and availability for crop growth vary considerably from soil to soil. Alternatively, their roots can grow into areas of a higher potential of soil water. But, as the store of water in the soil is tapped, the water potential falls; and because the plant cannot have a higher water potential than the soil with which it is in contact, the plant becomes more stressed. Again, since soils are so variable, both within and among areas, the only substitutes for the lack of the necessary experimental programmes are experienced, practical agronomists. However, peasant farmers are often the best judge of their own soils once they have gained experience, especially once a minimal amount of technical education has been offered.

Clearly, the effect of a given quantity of water on crop output will depend greatly on the timing and conditions at and after the application. Although the yield of useful product will be reduced by stress, it will not be economic to minimize stress because this could involve very large quantities of water very frequently applied. The irrigation system that gives maximum yield will maintain the entire root zone close to field capacity through continuous irrigation. Consequently, where water is unpriced, farmers will try to apply water frequently and in sufficient quantities to maintain full soil moisture.

From an economic point of view, four categories of stress can be recognised:

(1) stress which does not affect physiological processes;

(2) stress that has a temporary effect which can be overcome by subsequent compensatory growth;

(3) stress that reduces useful crop products;

(4) stress that results in crop death.

The first category is a common condition and is probably necessary for normal functioning of the plant. The second category occurs where stress is approximately between −5·0 and −10·0 bars, and results in a slowing of cell division and elongation. Stress beyond −10·0 bars causes stomata closure, and the price of the consequent saving on transpiration is an interruption of CO_2 exchange and thus of photosynthesis. Slatyer (*op. cit.*) suggests photosynthesis is affected with deficits close to zero (−1 to –3 bars) before cell enlargement and division but it is not clear whether it is likely to have economic significance. As to the third category, plant hormones which impede growth, and ultimately yield, may be produced by the plant under severe stress. This is a useful protective device, but it may take a plant up to a week after stress ends to metabolize these hormones and resume normal growth. Research work is currently under way which may discover chemical means to speed up the removal of growth-inhibiting substances as soon as stress is relieved.[6] It is conceivable that in future, with more basic scientific research, plants could be screened (using abscisic acid levels as an indicator) not only for their capacity to produce drought-control substances but for their speed in removing them. Alternatively, in future, farmers might apply sprays, immediately after rainfall or irrigation, of materials that destroy the now unwanted substances which keep stomata closed or otherwise act as growth inhibitors.

In addition, crops could be selected or bred for their capacity to occupy a large soil volume to capture available soil water; cultivars could be selected with a more equal distribution of roots within the root zone. Chemical as well as mechanical means might be devised to encourage soil penetration.[7] Plants might also be selected because of their low internal resistance to water movement; a capacity to translocate moisture rapidly would be a valuable attribute in desert environments. Leaf characteristics can be a focus for variety selection. It has been shown that cultivars of plants vary widely in their genetic capacity to withstand drought, and it is now evident that this is linked in a complex manner to endogenously generated plant-growth hormone inhibitors, particularly to abscisic acid, which affects stomatal closure.[8]

Hurd (1976)[9] reviews a variety of screening tests for drought resistance. Drought resistant crop varieties may appear to have little relevance for irrigation schemes but given the general unreliability of irrigation supplies in most developing countries this is clearly a desirable characteristic in many circumstances.

Factors affecting evaporation and transpiration

In discussing evaporation and transpiration, it is always necessary to distinguish between potential and actual water use. Potential evaporation and transpiration are determined by laws of physics related to energy exchange. However, other environmental factors often intervene which may invalidate a simple energy balance approach.

Water use in crop production is influenced by climatic, plant, soil, and cultural factors. The dominant climatic factors are solar energy input, wind speed, and humidity. Crop evapotranspiration or consumptive use is made up of plant transpiration to the atmosphere, plus evaporation from adjacent soil or other moist surfaces. It is possible to determine evapotranspiration indirectly by using meteorological data and mathematical procedures or by direct experiments.

There are at least six mathematical aids for determining crop evaporation. The most important are the water balance method (HMSO 1967),[10] combination of energy balance equations and aerodynamic relationships (Penman 1948),[11] and climatological methods (Thornthwaite 1948,[12] Blaney and Criddle 1950[13]). The methods are all empirical to varying degrees; each is a rational procedure of using available or obtainable data as an approximation for field or regional crop requirements.

A range of procedures has been used. At one extreme there are methods that use a theoretical base and climatological parameters such as temperature and wind speed, which can be accurately measured but which are remote from crop growth. At the other extreme are heavily empirical procedures that measure water inputs and crop outputs directly. These direct methods require data which are difficult to obtain in an accurate form, though they are clearly closer to practical application.

In irrigation practice theoretical approaches such as the Penman method find favour at the planning and design stage, with direct soil moisture measures being preferred for day-to-day operation. The climatological methods use air temperature as an index of energy available for evapotranspiration and work well in subtropical or temperate continental climates where temperature is a good index of net radiation. In arid areas and tropical zones Thornthwaite, Blaney and Criddle methods seriously underestimate the amplitude of seasonal fluctuations in water requirement and can give wrong indication of monthly water requirements. They should not be used for arid zone or tropical area irrigation planning. Methods based on Penman equations are thus preferable although radiation data (and other climatological measures) will be required.[14]

THEORIES OF WATER REQUIREMENTS

The old idea that each plant had a 'transpiration ratio', or given amount of dry matter which must be grown per unit of water transpired, has been abandoned by plant physiologists. It is now recognized that, for a given growth of dry matter, the amount of water transpired may vary greatly according to climate and circumstances. If the maximum temperature is not to be exceeded—that is to say, if the plants are to live—the cultivation must have a 'thermal balance'. This requires that the heat inflow from solar radiation, plus a small amount conducted upwards from the earth and sometimes from hot winds, is balanced by the heat outflow through radiation and convection, the effect of cooling winds, and heat stored in the soil to be disposed of at night. The balance of heat must, inescapably, be disposed of in the form of evaporation from the soil and transpiration from the plants. The factors in this thermal balance, it is clear, are very complex.

Thornthwaite devised an empirical system of predicting evaporation approximately from available climatic statistics. This has been checked satisfactorily against the known volume of evaporation of water from certain American irrigation areas, but it may not be universal applicability. Twelve valleys in the United States were studied with the figures for theoretical evaporation (averaged over the year) ranging from 1·2 to 3·2 mm/day. Actual figures were found to be within 4 per cent of the theoretical in all cases except West Tule Lake in California (actual figure 15 per cent in excess of theoretical). The Barohona Valley in the Dominican Republic, in a much hotter climate, showed a theoretical annual average of 4 mm/day, which was precisely confirmed in fact. Thornthwaitc's formulae, however, have not always proved satisfactory, and there have been others proposed. A formula has been suggested by Quesnel[15] to the effect that with solar radiation at thc rate of R calories/cm horizontal surface/day and average monthly temperature $T°$ centigrade then potential evapotranspiration in mm/month will be

$$\frac{0{\cdot}4T(R+50)}{(T+15)}$$

Other empirical measures for field-scale or larger-scale areas exist using radiation, humidity, or evaporation rather than temperature as the principal climatological parameter. Water-balance methods include studies of catchment hydrology, soil moisture depletion studies, and lysimetry. Lysimeters are instruments devised for direct measures of evapotranspiration. Plants are grown in weighed containers in the field, and water use is estimated from changes in weight. Lysimeters have often provided unreliable information because vegetative and soil-moisture conditions differed from those of the surrounding crop; effective leaf area for interception of energy and for transpiration is too large; and because of various edge effects stemming from the large proportional border area. Nevertheless, lysimeters can provide accurate measurement of evapotranspiration, especially in areas of high rainfall.

New instrumentation for measuring total soil water potential has recently become available which increases the measurement range and makes *in situ* measurements possible. This has great possibilities for future field application, not only in detecting deficit and indicating its extent, but also for control, for example, with automated sprinkler irrigation.[16]

It is clear that nearly all the factors determining evaporation are independent, or virtually independent, of the nature of the crop being grown.[17] The climatic factors—solar radiation, wind movements, and so on—are entirely independent. The temperature limits which have to be observed do not vary greatly between plants: some naturally occurring desert plants have a higher temperature limit, but they are not of agricultural importance. A significant but minor factor is that some plants (for example, sugar cane) have glossier leaves than others, and are therefore able to reflect back a little more of the sun's radiation, thus slightly reducing their water requirements. Other factors include different insulating properties of leaves, different effects upon turbulence, and different stomatal closure patterns.

Penman elaborated the principle of energy balance between incoming solar energy and wind energy and evapotranspiration. His researches, or put another way, his indication of an obvious fact of physics, which should have been recognized earlier, must necessarily lead to the drastic revision of all previous ideas about the economics of irrigation. Irrigating a given area at a given time of the year will use up the same amount of water almost irrespective of the crop which is being grown. Therefore, in principle, irrigators should always be growing the crops which, at that time of the year, and at the prices then prevailing, yield the highest economic return per unit of their scarce factor (be it land or water) and per unit of time. This statement is based on the assumption, which is the case with most irrigation farmers, that land and water are their constraining factors. In some cases, where labour is the limiting factor, a different policy is appropriate.

Non-climatic influences on growth

The influence of plant, soil, and cultural factors on plant growth, though generally much less important than climatic factors, has to be assessed. Plant characteristics which require modification of the Penman approach of assuming that all crops have equal requirements in given climatic conditions, include the number and distribution of stomata, the leaf texture and coatings, and various other areas of internal resistance to diffusion of water. However, knowledge of actual values over a wide range of conditions is still meagre, and there is great need for further research in this area.

Climatic and vegetational factors

Rutter,[18] in a model study using representative values for reflection factors, aerodynamic roughness, and internal plant resistance, investigated the interaction of climatic and vegetational characteristics in grass 10 cm high, a crop 100 cm high, and forest 5 m high. His exploration showed that the increased transpiration from forests that might occur because of their greater height and lower air resistance would be more than offset if their internal resistance was double that of grass and short crops. This is in fact supported by the estimates of internal plant resistance made by Holmgren *et al.* (1965) and Monteith (1965) that were quoted by Rutter. Rutter's model shows that transpiration from grass slightly exceeds that from forest, and transpiration from the tall crop exceeds both grass and forest. However, at higher levels of radiation (above the equivalent of 7·5 mm/day), a transpiration is greatest from the tall crop, and from forest is greater than from grass at low humidities (vapour pressure deficit > 4 m bar). This agrees with the empirical estimates of Monteith for a humid site in England (grass and forest having the same potential transpiration) and the arid Sacramento Valley (forest showing 23 per cent greater transpiration than grass).

The importance of this discussion for irrigation planning is that there is clearly an interaction between climatic and plant characteristics. This interaction and the importance of vegetation characteristics increase in extreme climates such as are found in arid zones.

Soil factors

The operation of irrigation, and the planning of the amount and frequency of application of water, depend on the effective volume of soil within range of the plants' roots (and the number of roots), the water-holding capacity of the soil, and the rate at which the soil surface can absorb water without it running off.

Soil moisture will often be available at different rates at various depths. Typically, the upper layers of the soil which contain most active roots and are subject to direct evaporation will be the first to dry out. Economists have often built models which regard the soil as being homogeneous by moisture, but Palmer-Jones[19] adapted a theoretical formulation by Gardner[20] for his study of tea irrigation in Malawi, in which he develops a model incorporating two soil layers where the density of roots differs. This is used to estimate relationships between yield, soil moisture conditions, and atmospheric environmental characteristics.

The 'effective depths' at which plant roots can extract moisture from the soil and other relevant data were stated by Molenaar[21] (for deep, well-drained soils) and are shown in table 2.1 below.

The distribution of roots in relation to moisture in the soil is equally important for crop nutrition. The major source of plant nutrients is the top few centimetres of soil. Often there can be a growth response to irrigation in a crop that does not show symptoms of stress. If a crop with a dry topsoil obtains water supplies from a depth but remains deficient in nutrients, then an irrigation which moistens the topsoil may facilitate nutrient movement and crop growth.

Table 2.1. Effective depths for plant roots, water-holding capacity, and infiltration rates (cm).

Onion, lettuce	30
Pasture, potato, bean, cabbage, spinach, strawberry	60
Sweet corn, table beet, peas, squash, carrot, eggplant, peppers	90
Sugar-beet, sweet potato, cotton, citrus, lima bean, artichoke	120
Melon, flax, maize, small grains	150
Alfalfa, asparagus, non-citrus orchard, grapes, hops, grains other than maize, sudan grass, sorghum, tomato	180

Soil type	*Water-holding capacity (mm/cm depth of soil)*	*Maximum rate of water intake (mm/hr)*
Very coarse sands	0·4	19–25·5
Sands	0·7	12·5–19
Sandy loams	1·05	12·5
Medium loams	1·6	10
Clay loams	1·75	7·5
Clays	1·70	—

This information suggests that availability of water is a function of the rooting depth, the water-holding capacity, and the rate of application. The effective storage capacity of a soil, therefore, instead of being a uniform 10 cm, as Thornthwaite originally proposed, would appear from the above figures to vary from 1·2 cm for lettuces grown in coarse, sandy soil to 315 cm for sorghum or tomatoes grown in a clay loam.[22] Therefore, lettuces grown on a sandy soil would require frequent light irrigations. Tomatoes grown on a clay soil would have considerable stored soil moisture to draw upon but, because of low infiltration rates, irrigation applications would have to be slow.

This idea, however, would be strongly controverted by Penman, who believes that each plant does not have a fixed rooting depth, but tends to adapt it in inverse proportion to the water-holding capacity of the soil, so that the effective storage capacity of all types of soils appears to be nearly uniform for a given crop (though no doubt he would except a few crops such as onion and potato, which seem to be always shallow-rooted).

On the whole E. W. Russell thinks the same:

> Frequent light irrigations encourage shallow rooting, and infrequent heavy irrigations, deep-rooting . . . because the roots can find all the water they need in the superficial layers of the soil . . . [and] plants are usually deeper-rooted in light sandy than in clay or loam soils.[23]

However, he does also mention the possibility that roots may be repelled by the higher CO_2 concentration at the deeper levels in clay or loam soils.

Russell, however, has added that some of the ill-effects which he has in the past blamed upon CO_2 appear, in fact, to have been due to a whole range of complex, harmful organic compounds which also accumulate in the deeper levels of heavy soils. Nevertheless, his conclusions concerning their harmful nature remain.[24]

In making the choice between frequent and infrequent irrigation, it should be borne in mind that it is probable (though there is not universal agreement on this point) that the rate of photosynthesis per unit of leaf area remains unaffected by temporary deficiencies leading to low levels of stress.

In specifying water requirements for plant growth, the important coefficient is the amount of energy which the plant's roots have to exert in order to extract water from the soil. Schofield measured this by the logarithm of the height in centimetres of the imputed capillary column of water required to effect this extraction.[25] This coefficient *pF* stands at 7 when the soil is oven-dry; under normal conditions of field capacity it may be about 2; a value of 4, or about 10 atmospheres, indicates the 'wilting point' of plants (with a few exceptions such as cacti). It appears that this is the maximum energy which the plant can summon up for the purpose of extracting water from the soil. What is important is the amount of water held at tension less than 4 on Schofield's scale.

Some soil scientists, however, hold that the wilting point occurs not through lack of energy on the part of the plant, but because water cannot travel rapidly enough through the zone adjacent to the roots. The first wilting point, namely the *pF* at which crops wilt during the hottest part of the day but recover at night, is perhaps that at which water movement through the soil is too slow to meet demands for transpiration. Ultimate wilting point is usually measured at the point at which the plant has become wholly and finally incapable of transpiration, even in a saturated atmosphere. The theory of the inability of water to travel fast enough through the soil may be criticized in the light of the consideration that pressure differences within the soil are small in comparison with the diffusion pressure between leaf and atmosphere. However, the hydraulic conductivity of soil is lower than the xylem (water transporting) vessels of the plant, which may account for the relatively slow movement of soil water.

Field capacity is not the same as saturation, when all the air spaces in the soil have been filled with water. It represents the state of affairs to be expected two or three days after saturation, when the forces of capillarity and gravity balance each other, and downward percolation has almost stopped. A deep-rooting crop such as lucerne will be less influenced by soil moisture conditions than a shallow-rooting crop such as grass. One main reason why run-off from forest is less than from grass catchments (contrary to what foresters would have us believe) is that trees go on drawing on deep soil water when soil moisture in surface layers is depleted. Grass comes under moisture stress at these times, leaving deep soil moisture unused. Rutter (*op. cit.*) quotes data which show that the evaporation of grass as a percentage of evaporation of adjacent forest is 52 to 100 per cent with a negligible soil moisture deficit, and between 60 and 92 per cent with various levels of soil moisture deficit.

The level of the water-table below ground surface is the most important factor that determines whether soil evaporation is likely to enable potential evapotranspiration to occur. The subsoil level from which water will move upward in the liquid phase at the potential evaporation rate varies with the soil moisture characteristics and soil conductivity. It is usually quite shallow. Lowering of the water-table below 2 metres usually precludes upward movement of soil moisture to the surface in the liquid phase. Empirical testing of this in the Lower Indus Project to determine an optimum level for stabilizing the water-table, showed that evaporation from the water-table was, in Indus alluvium, 100 per cent of potential at 60 cm but less than 10 per cent at 2 metres. Such facts are clearly important when design decisions relating to the depth of drains are being taken.

Plant roots cannot tolerate a saturated soil (except peculiar crops such as rice). Therefore, the depth to the water-table (or to a perched water-table, if a clay pan exists) determines the effective rooting depth. In a situation where water-table levels fluctuate widely, the mean value of water-table depth may not represent a guide to rooting depth for perennial crops. For example, it will not be possible to grow tree crops in areas adjacent to rice areas where the water-table is typically close to the surface during cultivation. However, if the water-table falls after rice, it may be possible to grow a seasonal deep-rooting crop. In parts of India and Pakistan it is traditional to grow crops such as gram, peas, and even wheat after rice, their roots obtaining moisture by following the capillary fringe of the falling water-table.

Cultural factors

Crop transpiration can be reduced by 40 per cent with a rise in leaf temperature of only 2–2·5 °C. This is because the factors of heat disposition are interdependent. A slowing down in transpiration results in increases in sensible heat dissipation and reduction of the leaf temperature (though leaf temperature would be higher than with optimum transpiration). Gale and Poljakoff-Mayber,[26] presenting evidence of this, conclude that a relatively large (40–50 per cent) reduction in transpiration would not be detrimental to the plant. Insofar as a reduced rate may, at times of rapid transpiration, prevent development of a serious water deficit, it could be beneficial.

Transpiration can be reduced by modification of the plant or of the plant's environment. The plant can be made to reflect more energy or increase its resistance to evaporation. In the arid zones, where most irrigation projects are located, energy input is typically more than three times that which satisfies photosynthesis (not for some crops as maize and sugar cane) and most of the energy, although contributing to heating the plant, is of too long a wave-length for photosynthesis. Kaolinite sprayed on to leaves has been found experimentally to decrease transpiration 22–28 per cent by increasing leaf reflectivity (Abou-Khaled *et al.* 1970).[27] Practical problems abound, including toxicity of reflective materials, adhesion to leaves, their stability and permeability to gases, and the negative effects of reflective coatings upon photosynthetic activity at periods of low light intensity (for example, in the morning and at evening). Research on reflective coatings is continuing, and it may yield practical results. We may yet see fields of white or silver wheat in future irrigation schemes.

However, more promising research is being undertaken on chemicals which induce stomata to close and on inert leaf coatings which reduce water loss. These materials may reduce transpiration more than photosynthesis. Optimum control of stomata opening using chemical inhibitors is, however, very difficult to achieve in practice. One experiment with bananas using an anti-transpirant spray reduced water abstraction from the soil by one-third (Gale *et al.* 1964).[28] There are a number of problems related to cost, non-specificity, short-term action, and toxicity but a number of potential gains. These materials could be used for slow release of mineral nutrients, insecticides, and fungicide and may form a physical barrier to entry of pest and diseases. Film-forming materials tested to reduce transpiration include cetyl alcohol, silicones, wax, latex, and plastics. No ideal material has yet been discovered.

Modification to the plant environment can be achieved by wind-breaks or enclosures. Physical or plant wind-breaks are of value in reducing transpiration where advection is an important source of energy. Indeed, in desert environments, wind-breaks may be essential to prevent sand abrasion. In areas sheltered from advection the photosynthesis transpiration ratio may be increased. The problems of wind-breaks relate to establishment and maintenance costs and, in the case of live wind-breaks, to competition for soil moisture with adjacent crops.

Enclosures such as transparent plastic structures and glasshouses are widely used to increase and control temperatures. In arid areas the main use is to conserve moisture. Enclosures give opportunity for close environmental control. For example, the regulation of CO_2 concentration can be achieved. Incidentally, increased CO_2 concentrations induce stomatal closure, which makes CO_2 the ideal anti-transpirant, lowering transpiration but increasing photosynthesis. However, enclosures consume substantial quantities of capital and, if heated or cooled, high levels of recurrent financial resources, and are therefore economic only in exceptional circumstances.

Estimated evapotranspiration

A survey of the literature on crop water requirements does not always reveal whether the data related to open-pan evaporation, open-water evaporation, potential evapotranspiration, actual

Table 2.2. Parameters used, and resulting estimation, of crop water requirements, Saudi Arabia.

	Jan	Feb	Mar	Apr	May	Jun	Jul	Aug	Sep	Oct	Nov	Dec
Mean air temperature (°C)	14·9	16·3	21·6	25·3	30·5	33·0	34·5	33·8	31·2	26·2	20·4	15·8
Relative humidity (%)	53·0	46·0	44·0	41·0	27·0	17·0	16·0	18·0	23·0	31·0	44·0	49·0
Ambient pressure (mb)	1015	1015	1015	1015	1015	1015	1015	1015	1015	1015	1015	1015
Sun hours (hrs)	9·0	9·5	9·4	10·0	10·8	11·5	11·9	11·6	11·1	10·5	9·8	9·0
Windspeed (km/day)	151	168	189	202	202	197	190	180	149	146	149	144
Pan evaporation (mm/day)	5·5	7·9	9·9	12·2	14·8	16·5	17·9	17·2	14·9	10·9	7·7	5·8
Evaporation (mm/day)	3·7	4·8	6·6	8·1	9·5	9·9	10·0	9·6	8·3	6·6	4·8	3·6
Adopted crop coefficients												
Spring wheat	1·05	1·00	0·85	—	—	—	—	—	—	—	0·60	0·85
Lucerne	0·85	0·90	0·95	1·00	1·05	1·05	1·10	1·00	0·95	0·85	0·60	0·75
Winter vegetables	1·05	1·20	1·10	0·90	—	—	—	—	—	—	0·45	0·65
Summer vegetables	—	—	—	0·45	0·75	0·85	0·85	0·80	0·60	—	—	—
Dates	1·10	1·20	1·05	0·95	0·85	0·85	0·80	0·80	1·00	1·00	1·00	1·15
Monthly rainfall, maximum and average values (mm/month)												
Maximum (mm/month)	79·8	24·5	29·5	77·5	4·5	0·0	0·0	0·0	0·0	0·0	32·0	21·3
Average (mm/month)	12·1	6·1	4·9	19·7	0·6	0·0	0·0	0·0	0·0	0·0	4·6	3·0
Farm water requirements												
Crop factor (weighted average)	1·00	1·03	0·91	0·84	0·90	0·95	0·98	0·90	0·83	0·85	0·58	0·78
Cropping pattern evapotranspiration mm/day	3·70	4·90	6·00	7·10	8·60	9·40	9·80	8·60	6·90	5·60	2·80	2·80
Field water requirement (mm/day)	5·30	7·00	8·60	10·10	12·30	13·40	14·00	12·30	9·90	8·00	4·00	4·00
Monthly watering rate for 10-hectare farm unit												
Intensity of cropping (%)	80	80	70	40	40	40	40	40	30	20	70	80
Water rate (m^3/day)	424	560	602	404	492	536	560	492	297	160	280	320

Source: Sir M. MacDonald & Partners, *Aflaj Plain agricultural improvement scheme,* p. 87, vol. 1, Development Potential, Report to Kingdom of Saudi Arabia, Riyadh, 1975

crop transpiration, or cropping pattern water requirement. There will obviously be great variation among these data for the same location. Table 2.2, p. 33, indicates the parameters used to estimate crop water requirement on an irrigation scheme in Saudi Arabia.

Open-pan evaporation is nearly always much higher than actual evaporation. The difference varies with the types of pan and the characteristics of the location. The main types of pan in use are the United States Weather Bureau (USWB) Class A pan, the British Standard, and the Australian pan. The pan coefficient (open water evaporation ÷ pan evaporation) is typically 0·7 for USWB and 0·9 for the other two. However, there is a wide range of evaporation, particularly for the British Standard pan (which is a square pan protruding 75 mm above the ground and thus susceptible to variation in evaporation depending upon wind direction).

In the case illustrated in table 2.2, pan evaporation was corrected to obtain estimates of potential evaporation, and crop consumptive use was obtained from a product of potential evaporation and 'crop coefficients'. These crop coefficients are empirically derived factors which represent the deviation from potential evapotranspiration as a result of limited ground cover in the early part of the growing season and the various botanical characteristics which effect evapotranspiration (height, colour, stomata behaviour, leaf glossiness, and so on). Cropping pattern evapotranspiration can then be determined. Rainfall adjustments are made and a field efficiency factor applied to contain field water requirement. Field efficiency expresses water delivered to the field as a proportion of water used directly by the crop, and it may be as high as 2·0 on badly managed farms. The field water requirement is then deflated by the designed intensity of land use to obtain the daily watering rate.

A few results are quoted in table 2.3 below for calculated potential evapotranspiration (that is, the amount of water which a crop covering the ground would transpire if kept adequately watered).

The estimates of potential evapotranspiration for West Pakistan are considerably lower than evaporation figures. Tree crops, and a few others in their earlier stages of growth, may leave a proportion of the land uncovered by vegetation, and this again may effect some reduction in water requirements, as the bare soil can be allowed to rise to a higher temperature than the growing plant. (In a damp soil, evaporation from bare soil may be greater than from plants.) Even this advantage, however, is not quite

Table 2.3. Estimated potential evapotranspiration (mm/day).

Month	*Hyderabad*[1]	*Multan*[1]	*Peshawar*[1]	*Trucial States*[2] *(Persian Gulf)*	*Hawaii*[3] *(Molokai)*
January	2·6	1·6	1·1	4·2	4·7
February	3·1	1·8	1·1	4·8	5·3
March	4·8	3·3	2·1	7·0	5·0
April	6·4	4·9	3·2	9·2	6·5
May	7·8	7·0	5·4		8·6
June	7·0	6·7	6·2		8·5
July	5·9	6·3	5·4	12·4	10·5
August	5·4	5·7	4·7	12·9	10·5
September	5·4	4·9	4·1	10·0	8·3
October	4·7	3·8	3·1	8·7	8·0
November	3·3	2·3	1·8	6·3	6·9
December	2·4	1·6	1·1	4·0	4·7

1. Revelle Commission Report. *Report on Land and Water Development on the Indus Plain,* US Government, 1964, p. 411.
2. Mohammed, *Man, Food and Agriculture in the Middle East* (American University of Beirut), p. 367.
3. *Agricultural Economic Report 72,* University of Hawaii Experimental Station.

what it seems; a plant surrounded by bare ground may transpire more because of advection of warm air from adjacent areas. For obvious reasons this is known as the 'oasis effect'. In arid areas this can dominate evapotranspiration. Not just individual plants but small projects in arid areas are open to this source of energy which creates increased evapotranspiration, thus illustrating some of the complications stemming from the interdependence of climatic variables.

Requirements less than the above are, however, quoted by Wynn[29] for the Sudan of 6·3 mm/day, even for the hottest months; less when there is any rain or cold. In the Gezira Irrigation Area of the Sudan, irrigation designed to meet potential evapotranspiration begins in August with an average input of 6·5 mm/day (including 2·5 mm average rainfall in that month). It is then reduced slightly and raised to 7·5 again in November, then maintained at about 7·5 until February, then lowered to 7 again before watering ceases in April.

A satisfactory measurement of evaporation should take into account the cooling effect of other evaporation which may be taking place in the neighbourhood. For this purpose, most evaporation pans are far too small. Ideally, evaporation should be measured in the middle of a large lake. We can, however, approach this ideal from the calculated aggregate evaporation from the huge Nile Swamps in Southern Sudan. The annual average is 6·1 mm/day, which appears to be substantially below that which would have been measured on the hot surrounding lands.

Transfer of heat between the soil, or crop, and the atmosphere is controlled by the differences of their temperatures and vapour pressures and the saturation vapour pressure of the atmosphere at current temperature, the velocity, temperature, and humidity of the winds, the roughness of the soil and the crop, and the crop's height above the ground. Most crops are able to reflect back about one-quarter of the incoming solar radiation, as against one-twentieth for an open water surface.[30] However, the reflection coefficient is not a constraint, its value depending upon the angle of the incoming radiation. (In the case of the Nile Swamps, partly covered with papyrus, the roughness of the leaves virtually offsets their reflection coefficient.)

Table 2.4. Evaporation or evapotranspiration, Sudan (mm/day).

	June	*July*	*August*	*September*	*October*
Evaporation from open ground	14	14	13	10	6
Evapotranspiration from rice field[1]			14–15	10	7·5
Evaporation from bare earth surrounded by rice			11	7	4

Source: Hunting Technical Services, private communication.

1. Judging by evaporation rates in the Sudan quoted earlier in this chapter, it seems likely that this rice field suffered from an oasis effect.

The point is further illustrated by observations in a rice field in the Sudan. The summer temperatures and evaporation must be among the highest in the world. Subject to this consideration, the bare soil, it is seen, can be allowed to rise to a higher temperature than the planted soil. The lower figures in the last line show that the rice plants near the soil are cooler than a plain open field. This is the reason why experiments conducted with single pot plants, with open space around them, do not reproduce the thermal conditions of the field, and overstate the plant's water requirements.

The complete unreliability of pot experiments, which fail to reproduce the thermal conditions of the open field, is illustrated by Bernard's experiments[31] on growing Bahia grass in the former Belgian Congo. When this grass was grown in pots it showed evaporation ranging from 3 to 9 mm/day, closely proportional to leaf area; but a year's lysimeter tests in the field showed a uniform evaporation of approximately 1·1 m (3 mm/day), whether growing naturally, watered, or watered and fertilized. In the two latter experiments the yields were raised to 44 and 63 tons/day matter/hectare/year respectively, without altering the water requirement significantly.

Wittig[32] quotes an experiment on maize in France, undertaken with the object of testing if water requirements depended on whether application was by furrow or by spray. No significant difference was found. In the case of spray irrigation, 65 per cent was evaporated, as against 15–20 per cent by furrow irrigation. It appears that spray irrigation performs much of its required function of cooling the plant simply by evaporation from the leaves and stems, without entering the root system.

Water requirements of various crops

Many irrigation engineers and administrators will still stoutly assert that the water requirements for a given area (or 'duty', as it is sometimes called in India) vary greatly according to the nature of the crop. Thus, for example, the Indian Ministry of Agriculture[33] states definitely the water requirements of rice at 1310 mm, of cotton at 280, of coarse grains (ragi, cholam, and cumbu) at 360, 220, and 170 respectively. Ghulam Mohammad[34] states requirements at 700–800 mm for rice, 350–450 mm for cotton. Sir M. MacDonald & Partners, in their Report No. 2 (1958) to the Government of Iraq on the Diyala and Middle Tigris Projects, first estimate water requirements for the different seasons of the year on the Blaney-Criddle system, but then apply 'crop factors' ranging from 0·6 for orchards and vegetables to 1·1 for rice. However, the effects of this differentiation are in the end cancelled out; total evaporation requirements (on the arable land, constituting some 85 per cent of the settled area) are estimated at a fairly steady rate ranging between 1100 and 1400 mm (annual rate), except for the November–February period, during which the rate ranges between 800 and 950 mm.

However, more careful examination shows—as indeed it must—that differences in water requirements arise from one of the following causes: (1) some crops have a longer growing period; (2) some crops are grown at seasons of the year when solar radiation is greater; (3) the extent of the minor factors, such as leaf reflection and bare soil, mentioned above; (4) the fact that, even when all the other factors are the same, some crops are usually grown on lighter soils, from which there may be a much greater loss of water by seepage than from others.

Where tall crops are grown so that they are exposed to drying winds, a 'clothes-line' effect can occur. Thus wind-break trees may have much higher water requirements than low-growing crops, and their evapotranspiration may exceed open-water evaporation. However, evapotranspiration is almost always less than open-water evaporation. In Hawaii,[35] evapotranspiration by cucumber and tomato was estimated at 60 per cent of that from an open-water surface. Evaporation, though, from grass in New Zealand was estimated[36] at only 20 per cent of pan evaporation. The Californian Department of Public Works conducted a series of lysimeter tests showing that evapotranspiration could range from 310 mm/year on bare earth to 660 for natural vegetation and 880 for willow plantations, to 1500 for an open-water surface. The figures for other crops shown in table 2.5 show considerable

Table 2.5. Evaporation, California (mm).

	Growing season	*Whole year*
Celery	360	460
Haricot beans	410	650
Potato	460	640
Onion	490	650
Grass pasture	660	660
Fruit	690	760
Sugarbeet	700	860
Market gardens	730	800
Wheat	740	880
Asparagus	820	820

Source: California Department of Public Works, *Bulletin* no. 27, 1931.

variations, which nevertheless may be capable of explanation in terms of lengths of growing season and of the proportions of the soil left bare. It is probably worth repeating that crop water requirements are determined by energy inputs and that, except for differences related to wind exposure, all crops in a given area receive the same energy in a particular period of time.

Total growing season requirements in western Oklahoma, including surface evaporation and unavoidable waste, were estimated as shown in table 2.6.

Where rainfall is reliable, irrigation requirements will be commensurately less. Gruner[37] considered that in Sicily, with an annual rainfall of about 700 mm, orchards needed 200 mm of irrigation, fodder crops 700–800 mm.

Table 2.6. Water requirements, western Oklahoma (mm).

Cotton	990
Lucerne	910
Bermuda grass	890
Wheat	810
Sorghum	740

Source: D. O. Anderson, N. R. Cook, and D. D. Badger, *Estimation of Irrigation Water Values in Western Oklahoma,* (Oklahoma State University, Stillwater, Okla, 1966)

Of all crops, rice is the most dependent upon irrigation, because the biology of the rice plant requires that the whole field should be waterlogged or actually under water during the planting season. Only under rare circumstances can this result be brought about by natural rainfall, and in most cases irrigation from streams or wells is necessary. The water requirements of rice have received more study than those of most irrigation crops because they are high and because of the great economic and social importance of the crop. In general, as will be seen below, our information about water consumption is still gravely inadequate. Average water requirements for rice in tropical climates have been estimated by the International Rice Commission at Bangkok at 1500 mm in all. These are the combined requirements for flooding the field at planting and for growing the crop, and are met by irrigation and rainfall during the growing season. A small amount (probably 200 mm maximum under the most favourable circumstances and with the most retentive soils) is receivable from water previously stored in the soil.

It may be added that the economic disadvantage of the high water requirements of rice is in part offset by the fact that the waterlogging discourages weed growth and makes available phosphates which would not have been available to dry land crops. The seepage from the flooded rice fields, moreover, may be re-used to perform the valuable function of removing accumulated salts in nearby basin-irrigation schemes (as in Egypt) on heavy clay. To achieve this object, ponding of water would be necessary in any case. In Pakistan it is normal practice to grow a crop of peas, wheat, or brassica oilseeds on the residual moisture after rice. Where the groundwater is fresh, seepage from the fields will recharge the aquifer, and it may be recovered and used in periods of scarcity when water has a high value.

Crop water needs may be much higher in clear, dry climates where the day temperatures may be exceptional and where there may be drying winds. Annual water requirements for rice have

Table 2.7. Plants' supposed water requirements.

Crop	*Country*	*Growing season (months)*	*Supposed water requirements (mm/day)*	*Do. net (from lysimeter tests)*
Bananas	Israel	12	5·7	—
Orchards	Iraq	12	4·7	—
	Israel	12	3·7	—
	Italy	12	2·3	—
	Pakistan N.W. Frontier	12	2·5	—
Lucerne and pasture	Israel	12	4·7	—
	Australia	12	4·3	—
	Italy (lucerne)	6	5·0	—
	Italy (grass)	9	3·3	—
	Australia (lucerne)	8	4·3	—
	Australia	7	3·0	—
Forest	Pakistan Punjab	12	3·0	—
Sugar-cane	Pakistan	11	4·0	—
	Pakistan N.W. Frontier	11	4·7	—
	Pakistan Punjab	11	4·3	4·0
	Pakistan Punjab	11	4·0	3·5
	Hyderabad	12	6·7	—
Cotton	Pakistan	7	4·7	—
	Israel	8	5·3	—
	Iraq	7	6·7	—
	Pakistan Punjab	7	4·2	4·7
	Pakistan Punjab	7	4·7	4·0
	Hyderabad	6½	5·3	—
Rice	General (see above)	7	7·2	—
	Pakistan	7	5·0	—
	Pakistan Punjab	7	9·0	6·3
	Pakistan Punjab	7	9·7	4·7
	Pakistan Punjab	7	7·0	5·3
	Hyderabad	3¼	11·0	—
Wheat	Pakistan	6	2·0	—
	Pakistan N.W. Frontier	8	2·7	—
	Iraq (including barley)	6	3·7	—
	Pakistan Punjab	6	1·3	1·7
	Pakistan Punjab	6	2·3	2·3
	Pakistan Punjab	6	1·2	1·3
	Hyderabad	3	4·3	—
Sugar-beet	Israel	6	5·0	—
	Italy	3	4·0	—
Fodder crops	Iraq (winter berseem)	6	4·7	—
	Iraq (summer)	5	8·3	—
	Australia (summer)	6	2·7	—
	Pakistan N.W. Frontier (summer)	4	5·1	—
	Israel (winter berseem)	3½	8·3	—
	Pakistan Punjab (winter berseem)	3	5·3	4·0
	Pakistan Punjab (winter pigeon pea)	3	3·0	3·0
	Pakistan (pigeon pea)	3	6·0	—
	Italy (clover)	3	5·3	—

Crop	*Country*	*Growing season (months)*	*Supposed water requirements (mm/day)*	*Do. net (from lysimeter tests)*
Maize	Israel	6	4·7	—
	Italy	3	4·8	—
	Pakistan	6	3·3	—
	Pakistan N.W. Frontier	6	3·7	—
	Pakistan Punjab	6	2·7	2·7
	Pakistan Punjab	6	3·0	2·0
	Hyderabad	3¼	3·7	—
Tobacco	Israel	5	6·3	—
	Australia	4	9·0	—
	Hyderabad	4½	7·7	—
Tomatoes	Israel	5	12·0	—
	Italy	3	5·7	—
Ground-nuts	Israel	4½	8·3	—
Millet	Iraq	3	6·3	—
	Pakistan	3	6·0	—
Melons	Israel	3½	9·0	—
Vegetables	Israel	3	11·0	—
	Italy	4	3·2	—
Potatoes	Israel	3	5·7	—
	Italy	3	10·0	—
	Hyderabad	3	7·0	—
Early potatoes	Israel	2	5·0	—
Barley	Pakistan	6	2·0	—
	Hyderabad	3	4·0	—
Sorghum	Pakistan	6	3·3	—
	Italy	3	4·7	—
Oilseed	Pakistan	6	2·0	—
Beans	Italy	3	3·3	—
Strawberries	Italy	5	4·3	—
Artichokes	Italy	6	3·8	—
Oats	Hyderabad	3	4·0	—
Vines	Australia	12	2·4	
Citrus	Australia	7	2·4	—
Deciduous fruit	Australia	7	2·3	—

Sources:

Australia: Queensland Irrigation & Water Supply Commission, Mareeba–Dimbulah Irrigation Project, 1952

Iraq: Binnie, Deacon & Gourley, *Report to Government of Iraq, Zad Irrigation Project,* vol. 1; Sir M. MacDonald & Partners, Report to Government of Iraq, Diyala & Middle Tigris Project, 1958

Israel: Stedman Davies, Middle East Agricultural Development Conference 1944

Italy: (Emilia-Romagná) Perdisa, quoted in Tofani, *Genio Rurale,* August 1955

Pakistan: R. Revelle, *Report on Land and Water Development in the Indus Plain,* US Government, 1964

Pakistan: *Crop Responses to Water,* Commonwealth Agricultural Bureau, 1967

Pakistan Punjab: Director of Agriculture, private communication, 1952

India: Director of Irrigation, private communication, 1952

been estimated at 1250 mm in Central Asia and 1800 mm in the Murrumbidgee irrigation area in Australia (where excessive use was also encouraged by an exceptionally favourable price). Water requirements in the dry, clear, long days in the summers of Spain and Italy also seem to be higher than in the humid, cloudy tropics. (We must also take account of the possibility, however, that the soils in question in Central Asia and Australia are more porous than those usually used for rice growing.)

A figure of 1900 mm was also estimated for rice growing in Iraq, but the general estimate of about 1500 mm was confirmed in two provinces of Pakistan. The experiment in India, quoted below, showed that *ad lib.* watering of rice, with total supplies up to 3500–4000 mm, did not increase yields in comparison with control plots with a supply of 1600 mm (in 5-day waterings). In many parts of the Indian sub-continent, water is run into rice fields continuously by farmers who believe that this will lower water temperature to the benefit of the rice crop. It is not known to the authors if this belief is valid but it certainly leads to very high levels of water consumption.

In table 2.7, the length of growing season (known or estimated) is compared with statements of supposed water requirements by a number of authorities (these latter sometimes omit the natural rainfall, which has then had to be estimated).

The effects of seepage can be very serious, particularly in rice-growing areas. Some farmers in Australia, visited by Colin Clark, had farms situated over beds of sandy sub-soil, and estimated that they were losing between 9 and 18 mm irrigation water per day by seepage. Even in the case of farms not so situated the loss was estimated at 2·5 mm/day.

There are a number of cases of serious over-watering (although, perhaps a tendency to under-water maize and wheat). Otherwise, the monthly figures (adjusted in some cases after lysimeter tests) appear on the average to be more or less what was to have been expected, in view of varying temperatures. A serious case of over-watering is reported from the Indian state of Maharashtra, where sugar-cane (crop cycle 18 months) receives an average of 320 mm/month (including 100–130 mm rainfall).[38] However, the most usual cause of over-watering is unreliability in the distribution system. This forces farmers to take all the water available to bring the soil profile to field capacity to ensure that they can survive any shortfall in supply in the near future.

In Italy, crops grown in the hotter southern region (Metaponto) may use less total water than in the north because of the short growing season. Likewise, in Australia in a hot region, total water requirements (including the rainfall of 23 cm) were estimated at 380 mm for cotton, 280 mm for vegatables, and for bananas 410 mm.

Error in estimation of water requirements

The tendency in India to over-water both rice and sugar-cane persists. See, for example, Indian official proposals for use of water from the large and costly Tungabhadra Dam (on the borders of Madras and Hyderabad).[39] The figures in table 2.8 refer to gross inputs—it is estimated that 25 per cent is lost between distributor head and the field—but even so they are clearly excessive.

Table 2.8. Proposed water inputs (mm/day).

Ground nut		4·0
Fruit, lucerne	March–June	6·0
	June–March	5·0
Vegetables	June–July	6·0
	August–February	4·5
Cotton	Mid-April–mid-May	6·0
	Mid-May–mid-June	5·0
	June–September	2·0
	September–October	1·5
Winter crops	Mid-September–mid-October	2·5
	Mid-October–mid-November	3·3
	Mid-November–mid-December	5·0
	January	4·7
Rice (tabi variety)	August–January	10·0
(abi variety)	June–November	9·0
Sugar-cane	April–June	9·0
	June–August	7·5
	August–November	9·0
	November–January	13·3
	January–March	10·3
	March–April	9·3

Irrigation planning procedures are in some dimensions over-elaborate and in some, too, superficial. In the final chapter it will be argued that many aspects of economic appraisal are excessively complicated. In the view of the authors, the estimation of water requirement for crops, which may or may not be grown depending upon relative market prices, could also be much more approximate. The current practice is to determine in a precise way monthly, or even every 10 to 15 days, water requirements for selected crops, often after field experiments. This water requirement is aggregated by assumed crop mixes; some notion of water-use efficiency is applied; some degree of leaching is prescribed; and water requirements are determined. Much of this information is uncertain: crop mixes may vary; new varieties may be developed with, say, a shorter growing season; irrigation efficiency is likely to be unknown, very different in practice from the assumptions, and to vary over the year and between years (improving, it is to be hoped, with time); leaching requirement will vary over time as the quality of water changes, in particular drainage is developed and drainage water is mixed with irrigation water, sometimes from projects upstream; farmers may cultivate a smaller area than predicted or, as is usual when water not land is limiting, they may spread the water to a larger crop area.

Even within the evapotranspiration calculations themselves there are likely to be significant errors. There will be data errors and omissions among the meteorological observations, and there is no agreement on an accurate or appropriate assumption of the probability of rainfall. More important, the various ways of estimating evapotranspiration, all of which have distinguished scientific supporters, give markedly different estimates. This is illustrated in table 2.9.

Misconceptions of the role of water in plants continue to arise. M. F. Ali[40] in discussing the Ganges-Kobadak project reports errors in design of pumps because 'they were based on the low-water consuming crops cultivated in the "fifties" ' and reports that the new high yielding cultivars require more water. We have noted that water requirements are largely determined by incoming energy. In the case of the high yielding cultivars they will receive the same energy (or to the extent they are short-duration varieties, less) and therefore will have the same water requirements.

Because of the variety and magnitude of errors that are possible in estimating water requirements for irrigation projects, as well as the probability of their normal distribution, the authors conclude that it is at the planning stage that a much more flexible approach must be taken. Potential evaporation is likely to be in the range of 1–3 mm/day in temperate areas, 5–8 mm/day in humid tropics, and 9–12 mm/day in arid zones. Such approximations could be modified in the light of local micro-climate, topography, and the season being considered. A 'model' crop mix with at most four crops—for example, for north India and Pakistan, cotton, sorghum, wheat, and oilseeds—might be used to determine monthly cropping pattern requirements with sufficient accuracy for engineering planners.

CROP RESPONSE TO IRRIGATION

The effects on gross physical product of increasing or decreasing water supplies will now be considered. Economic returns to water inputs, after all cost factors have been taken into account, will be considered in chapter 6.

Irrigation planning all too often ignores important basic economic concepts. For example, a recent authoritative text on irrigation defines optimum water requirements as the amounts of water required during the growing season to produce maximum yields of different crops.[41] This definition is at best ambiguous, but to many readers it will be interpreted as water sufficient to produce maximum yield per hectare. This interpretation is, of course, likely to be wrong because it is maximum return per unit of scarce input that matters, which may or may not be land and which, in most instances, is water.

The first step in establishing crop response is the simple analysis of the effects of changes in the total water supply during the growing season. The effects of the timing of water inputs, and the design of an optimum watering strategy, constitute much more complex problems. A recent excellent booklet has been published by the International Rice Research Institute on the practical problems facing field researchers attempting to establish by field survey the reasons for a divergence between potential irrigated production and actual levels of production.[42]

Irrigation inputs and crop yields

In fig. 2.1, three conceptions of seasonal water response curves are illustrated. The solid curve (*i*) is the classic response curve to a variable input showing zones of increasing returns (*o*, *x*), diminishing returns (*x*, *y*), and negative returns (>*y*). For economists, attention is mostly focused on the second zone, where it is likely that optimum response lies, the precise point depending upon the price of inputs and outputs and the alternative opportunities available. Many agriculturists have their eyes firmly fixed upon the input level that gives maximum yield (*o*, *y*), and believe that this coincides with maximum evapotranspiration. In the first zone, the physical return to additional water (marginal product) is greater than average product, and so the average product per unit of water is increasing. In the second zone, the marginal product is positive but less than average product. Once the maximum yield is achieved, the marginal product is zero, and any additional inputs will give a negative response. Curve (*ii*) is a

Table 2.9. Estimate of potential evapotranspiration for selected stations in West Pakistan (mm).

Method of Estimation	*Jan.*	*Feb.*	*March*	*April*	*May*	*June*	*July*	*Aug.*	*Sept.*	*Oct.*	*Nov.*	*Dec.*	*Annual*
Lahore													
1. Kohler (H–W–W)	50	59	116	170	243	224	176	160	166	137	78	46	1625
2. Kohler (Brunt)	40	46	97	148	219	202	159	140	147	126	68	38	1430
3. Penman	27	37	83	117	183	190	170	153	125	90	40	25	1242
4. Thornthwaite (Ahmad)													1372
5. Rohwer (Raman)	29	37	87	166	260	257	168	122	119	112	52	33	1442
6. Hiatt	41	48	94	142	208	193	160	142	132	114	66	38	1379
Quetta													
1. Kohler (H–W–W)	37	43	81	129	187	205	217	195	161	120	73	43	1485
2. Kohler (Brunt)	28	33	65	106	163	185	189	175	149	108	62	36	1299
3. Penman	22	29	63	95	152	169	177	157	106	72	34	23	1101
4. Thornthwaite	6	10	37	66	103	131	144	132	85	46	21	11	792
5. Rohwer	25	31	60	94	145	180	168	141	114	80	51	31	1121
6. Hiatt	30	36	66	107	157	175	180	168	139	101	61	36	1245
Hyderabad													
1. Kohler (H–W–W)	94	115	178	228	272	241	196	185	195	181	127	99	2109
2. Kohler (Brunt)	83	97	148	202	250	211	167	158	173	166	109	83	1844
3. Penman	54	65	120	161	239	227	192	180	157	129	67	52	1642
4. Thornthwaite													1549
5. Rohwer	115	113	217	301	419	396	335	260	232	208	151	125	2873
6. Hiatt	79	93	147	193	236	213	178	165	165	142	102	74	1788

Source and *Notes:* from W. E. Hiatt, 'Potential evapotranspiration estimates for West Pakistan', Appendix A.1 in R. Revelle. *Report on Land and Water Development in the Indus Plain,* US Government, 1964.

Method 1: Evaporation computed from the meteorological factors of air temperature, dew point, wind movement, and solar radiation, as described in Weather Bureau Research Paper no. 38, 'Evaporation from Pans and Lakes', by Kohler, Nordenson, and Fox (US Department of Commerce, Washington, DC)

Solar radiation estimated using relation shown in paper 'Insolation as an empirical function of daily sunshine duration', by Hamon, Weiss, and Wilson (*Monthly Weather Review*, **83**, no. 6, 141–46, June 1954).

Method 2: Evaporation computed in same way as Method 1. Solar radiation estimated using Brunt's equation.

Method 3: Evaporation computed using Penman's original equation, as described in *Netherlands Journal of Agricultural Science*, **4,** no. 1, February 1956, and other papers.

The computations by methods 1, 2 and 3 are based on average monthly meteorological data obtained from *Climatological Tables of Observations in India, 1953*. The averages are for the period 1881–1940, except stations Bahawalpur and Khanpur which are for the period 1926–40.

Method 4: Thornthwaite values of potential evapotranspiration were obtained from a map contained in report 'Precipitations, evapotranspiration et aridité en Pakistan Occidental', by Mohammad Shafi Ahmad. (No. 1, Trevaux du Laboratoire de Climatologie de la Faculté des Lettres, Université de Rennes, 1958.)

Method 5: Evaporation values computed by Rohwer's equation were obtained from paper 'Evaporation in India calculated from other meteorological factors', by P. K. Raman, published in *Scientific Notes*, India Meteorological Department, **4,** no. 61, 1935. The monthly values given in this paper were multiplied by 0·771, as recommended by Rohwer, to make formula applicable to reservoir evaporation.

Method 6: W. E. Hiatt, no explanation of basis.

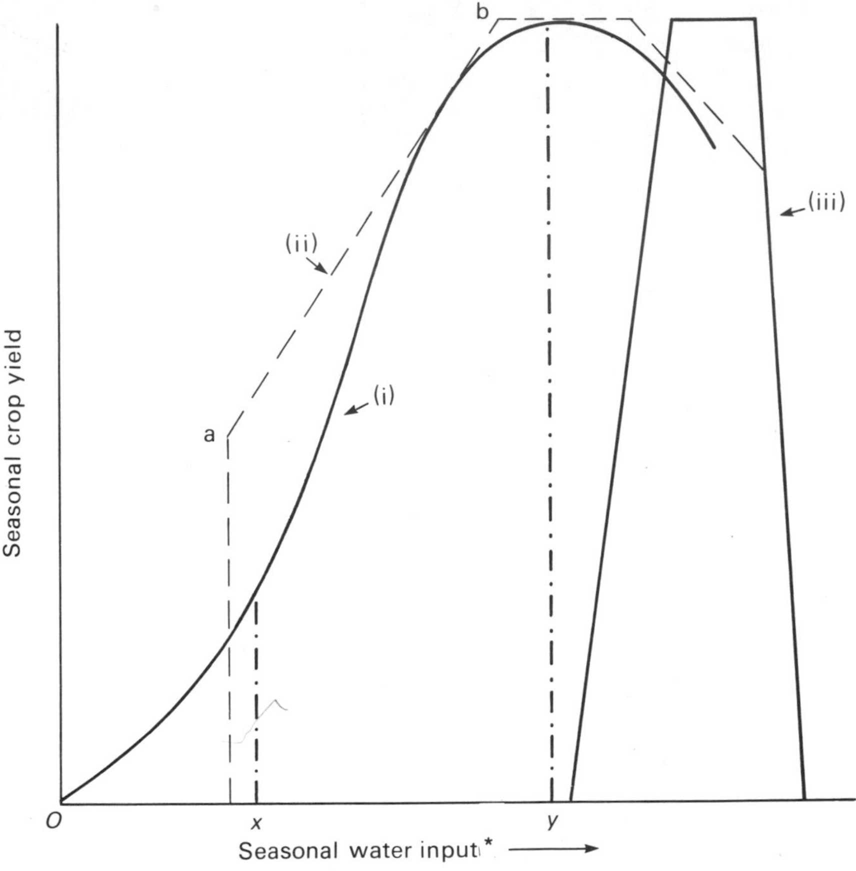

Figure 2.1. Yield response to water inputs.

*Conceptually, this could be water input over any time period—a month, week, or less, or a particular period in the plant's life cycle, e.g. flowering.

simplified presentation exhibiting linear response with increasing, constant, and then decreasing total returns per hectare from increasing water inputs. Curve (*iii*) might be expected to apply to rice with no yield obtained before soil waterlogging is achieved and a narrow range over which response occurs and a range where yield is constant before water inputs become detrimental. All three conceptions are gross simplifications, the most serious practical limitation being that, by using a seasonal input, the precise timing of inputs of water are assumed to be unimportant. Thus conceptually there is no effect upon yield levels if all the water is applied at once or as a constant trickle over the season, although in practice yield levels are clearly very sensitive to the application regime.

The treatment of time of application has become increasingly sophisticated. As yet no method is very satisfactory, and the data for estimating the production relations of water applied at different times is not readily available (see note 41). Economists have played an important role in emphasizing the importance of time, starting with Moore, 1961.[43] Mapp *et al.*, 1975, and Hexem and Heady provide recent examples.[44]

Despite the limitations of these simplified conceptions of crop response to water, for crops other than rice, where application regimes are similar, insights can be gained from analysis of field survey and experimental data that can help both to explain farmers' practices and to guide planners to improved design of projects.

The simplest conception of crop response to water is the linear response to seasonal inputs, and data relating to that portion of the response function exhibiting a positive return to increased inputs is typically examined (*a* to *b* on curve (*ii*) in fig. 2.1). Reutlinger and Seagraves,[45] in a pioneering study on sandy soils in North Carolina, showed that yields of tobacco (a shallow-rooting crop) fell more or less linearly from 2300 to 1500 kg/hectare in response to changes in the average over the whole growing season of a 'soil moisture deficiency'. Wisner's[46] method was selected, which assumes, in effect, that water does not redistribute itself between soil layers when the water content of the soil is below field capacity.

Approximately linear responses were also found by Beringer[47] to 'integrated soil moisture tension' (for which a formula was given) throughout the plant's growth. He concluded that a given total water supply, the application of which was distributed frequently over the growing period, should give a higher yield than the same water supply with less frequent distribution.

Field experiments in Pakistan showed a relatively low response to water, and significantly lower total and marginal returns again under actual farming as opposed to experiment station conditions.[48] The existence of a wide gap between experimental and on-farm levels of achievement is all too often ignored in planning.

Barker, analysing village data for rice response to fertilizer in six Asian countries, found a similar flat response curve and a big gap between research station response and that on farms. He also found a high positive correlation between average rice yields and the average level of applied nitrogen. There was no indication

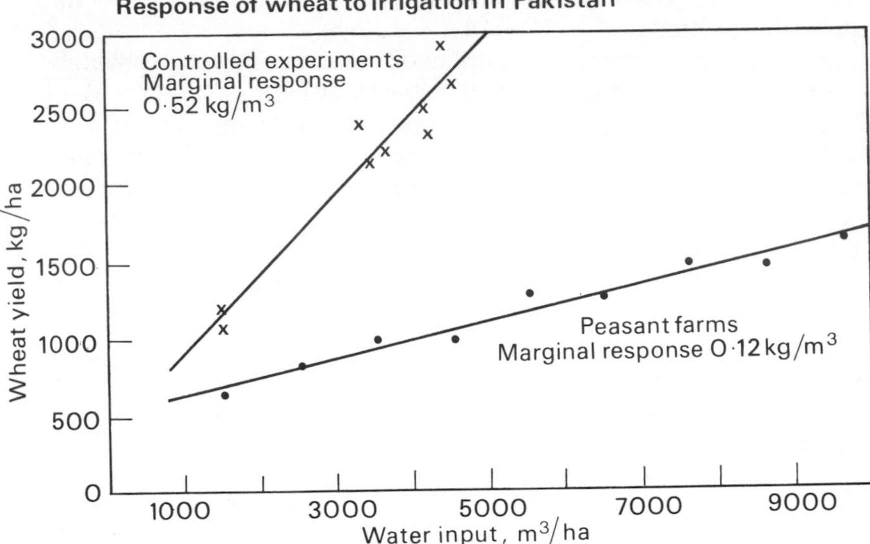

Figure 2.2. Response of wheat to irrigation in Pakistan.

that farmers could increase their (highly variable) average yield by increased fertilizer application.[49]

Very much higher marginal returns to irrigation water under experimental conditions than the 0·52 Kg/m^3 in Pakistan are reported from Israel. With one exception, marginal returns for wheat were between 0·76 and 1·60 kg/m^3, with most results more than 1·00 kg/m^3. These results stem from a review by Shalhevet *et al.* of experiments in several locations in Israel;[50] these writers give details of irrigation experiments to determine the 'optimum irrigation regime' for a variety of crops. There are, however, conceptual limitations in the design of many of these experiments, stemming from a failure to appreciate adequately the need to treat each irrigation application as a separate input (Palmer-Jones, 1976 and 1977).[51]

Guinet[52] finds yields of maize and potatoes to be linear functions of the difference between rainfall and potential evapotranspiration. Stewart[53] relates yields of irrigated maize linearly to actual transpiration. Averaging his results for 1970 and 1971.

Yield (kg/ha) = 189 (seasonal evapotranspiration in cm) − 2682
i.e., a marginal return of 1·89 kg/m^3.

Bell[54] considers that in low-rainfall country in Queensland (for which he recommends summer fallowing, even though maximum soil storage is only estimated at 150 mm), wheat and sorghum yields are determined by the amount of water (including stored water) in the top 900 mm of soil. Crops root deeply when water is scarce, and yield at the rate of 1·5 kg/m^3 of water.

In the case of grass in cool climates, marginal returns in physical units are best expressed per day (g/m^2/day dry weight) as annual results are affected by the low rate of growth in the winter. In New Zealand,[55] on soils, field capacity of which was 27–30 per cent and wilting point 10–11 per cent (volume measurements applied to the top 10 cm of soil, regarded as the limit for grass roots), the yields obtained for seasons of 91 days each are shown in table 2.10.

Table 2.10. Grass yields (g/m^2/day dry weight) on irrigated and unirrigated plots.

	Spring	*Summer*	*Autumn*
Unirrigated	2·8	1·1	0·8
Soil moisture maintained at:			
10–11 per cent	4·2	4·0	1·9
20 per cent	5·3	5·1	2·5

It is clear that the natural rainfall often leaves the top 10 cm, on which shallow-rooted grasses depend, below wilting point.

Richard and Fitzgerald,[56] of the same research station, estimate from a linear response function marginal returns of grass to varying water inputs (range 350–500 mm) in the September–May season.

Dry weight yield kg/ha = 183 (cms input) − 2710
i.e., a marginal return of 1·83 kg/m^3.

The same authors[57] consider that the irrigation yields can be kept in the 4–5 g/m^2/day range for all months in which average temperatures exceed 10 °C. In the warmer and more humid New Zealand North Island climate,[58] average yields over a 9·5-month season were estimated at 5·5 g/m^2/day, but still capable of being raised to 6·2 by irrigation; this latter yield in the form of grass and silage would feed 5 cows/ha without any supplementation.

On irrigated pastures in a hot climate at Yanco in Australia, the average yield is given[59] as 5 g/m^2/day but the maximum as 10. With really thorough watering and fertilization, with cow manure, in a hot climate, this figure can be raised[60] to 16 (60 tons dry weight/ha/year, which comes close to the maximum of 75 tons obtained in experiments in the tropics). The Ruakua Institute estimated a marginal return to pasture of 1·6–1·8 kg/m^3 dry weight, keeping cattle at optimum density.

Limitations of efficacy of water application

It is somewhat surprising, in the light of results, that so many experiments on crop response cover a very narrow range of applications. For example, none of the treatments in the controlled experiments illustrated in fig. 2.2 above reached potential evapotranspiration (6000 m^3/ha). Furthermore, despite a lack of experimental evidence that this level of water application produced maximum yield per hectare, all the irrigation schemes in the Lower Indus Project were designed as if farmers would use such amounts, and agricultural extension officers advise such applications. Only 15 per cent of farmers in a survey of existing schemes were in fact applying to these levels, 44 per cent applied less than 3500 m^3/ha. Given the excess of land compared to water, they were quite rational to do so.[61] As Falcon and Gotsch point out,[62] where land is unused and virtually free (as it is in many low-rainfall areas such as the Indus Basin), it pays the farmer to apply water only up to the point where its *marginal return* is at its maximum. Given a response function with diminishing returns (or a linear function with a positive constant), a rational farmer will indulge in what appears to be, from a technical viewpoint, under-watering. However, it should be noted that in arid areas spreading water to obtain maximum returns per unit of water will, in time, lead to a serious problem of soil salination (to be dealt with in chapter 3).

A measure of the marginal response of a crop to irrigation can provide the basis for a useful first check on the economic returns to irrigation. This is true even when it is derived from a linear function, providing that the limitations, relating to the narrow range of the response curve where the information is valid and the other simplifying assumptions, are borne in mind. The phenomenon of diminishing marginal returns is the norm for crop response to variable factor inputs such as fertilizer or water. An indication of this can be obtained while retaining the simplifying linear assumptions by looking separately at segments of the response curve. For example, in Bangladesh, a high rainfall area but with dry weather in the winter, the yield of winter rice is greatly improved by the addition of the first 340 mm of irrigation water, the return (in terms of rough rice) being 0·56 kg/m^3. Further watering to the extent of an additional 700 mm, however, yields additional rice at only one-tenth of that rate.[63]

Irrigation farmers generally recognize as well as or better than planners the existence of diminishing returns to increments of water per hectare, but in order for them to react appropriately they have to face the real costs of the water provision and have access to a reliable supply.

Gotsch makes the point that the Pakistan system of allotting water in rotation, rather than selling at specified prices to buyers when they need it, leads to a great deal of waste, and even in some cases to over-watering, to the detriment of crops.[64] However, this waste can be exaggerated, for as the data illustrated in fig. 2.2 indicated, the practice of under-watering is more prevalent than over-watering. Furthermore, it is the unreliable system with a risk of non-delivery on the expected date (due mainly to the failure of the irrigation authorities to prevent illegal upstream abstractions) that leads to most of the over-watering that does occur. By and large, in the Indus system farmers choose to cultivate a larger area than they might reasonably expect to bring to harvest, in the hope that, fortuitously (or illegally), additional supplies will be forthcoming.

Intuitively, it is clear that a Gompertz or quadratic form of function is preferable to a linear form when a wide range of water inputs is being considered.

It should be clear there must be limits to linear responses. Experiments at the Wellesbourne Vegetable Research Station[65] in England show that, with a rainfall of 610 mm, the yield of peas ceases to increase after an additional 250 mm input of water, but with cauliflower and potatoes it goes on increasing up to 690 and 740 mm total input respectively. Robelin[66] performed an interesting series of experiments in which widely differing rates of evapotranspiration were obtained, partly by watering, partly by choice of soils. In one set of experiments the crops were watered heavily enough to bring the water-table to within 600 mm of the surface. This had the effect of reducing the supply of oxygen and nutrients to the roots, so the results of these experiments were omitted; but apparently the roots of the lucerne were unaffected. Lucerne yields could therefore be measured over a wider range of water inputs, and show a definite slackening of increase above an input of about 800 mm. The lucerne results can be closely fitted by a Gompertz curve as shown in fig. 2.3. On the other hand, the returns to maize and sugar beet, over the range studied, were linear. The experiments were performed at Clermont-Ferrand, in the Central Plateau of France, with hot summers and cold winters.

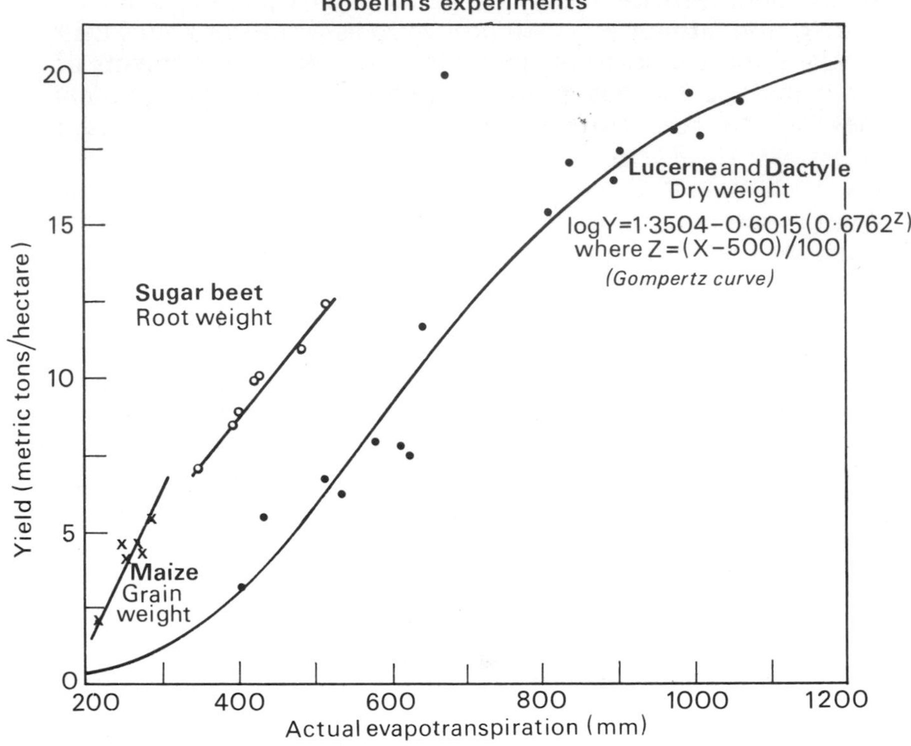

Figure 2.3. Yield response in Robelin's experiments.

Yaron[67] explored the data from experiments in Israel on cotton, sorghum, and ground-nuts with other functional forms and a combination of factors affecting yield. He found no greater explanatory power for yield when using a function including depth of soil moistening, frequency of irrigation, date of last irrigation, as well as total quantity of water applied. When he analysed data fitted by least squares to functions such as

$$y = b_0 + b_1x_1 + b_2x_1^2 \qquad (i)$$

and

$$y = b_0 + b_1x_1 + b_2x_1^3 \qquad (ii)$$

he found that curves for a given crop in the same location but for different years tended to run parallel to each other. Differences in elevation he hypothesized were due, partly, to differences in soil fertility between plots, but mainly to effects specific to a particular year, such as weather, pest, and disease. This 'year effect' hypothesis, if valid, is important in that the slope of the curves, and thus the marginal returns, do not vary between years, which makes it possible to use data from a series of experiments over several years to predict response to variations in water applications. The marginal returns that he established using equations (*i*) and (*ii*) are shown in table 2.11. With the range of data available it was not possible to choose between the two functional forms by statistical tests.

Table 2.11. Marginal returns from experiments in Israel.

	Marginal Returns					
Water application (m^3/ha)	*Cotton (2 yrs' data) (kg cotton fibre/m^3)*		*Groundnuts (3 yrs' data) (pods kg/m^3)*		*Sorghum (3 yrs' data) (kg grain/m^3)*	
	Equation (i)	*Equation (ii)*	*Equation (i)*	*Equation (ii)*	*Equation (i)*	*Equation (ii)*
1000	0·42	0·38			2·0	1·9
2000	0·25	0·34	1·24	1·07	1·6	1·7
3000	0·22	0·23	0·93	0·91	1·2	1·2
4000	0·20	0·18	0·62	0·65	0·7	0·8
5000	0·17	0·12	0·31	0·33		
6000	0·14	0·14				
7000	0·11	0·10				
8000	0·08	0·08				

Source: see reference 67.

There must be limits beyond which more water cannot usefully be added, and a point at which plants in fact drown. It is useful, however, to have this confirmed by the results (see table 2.12 below), of two experiments in deliberate over-watering, on rice in India and lucerne in the United States.

Timing of water application

Whether a linear response or some curvilinear form is assumed, the timing of the water inputs over the season is clearly relevant. Reutlinger and Seagraves (quoted above) considered that no adverse effect on plant growth would be felt until soil moisture had fallen to the halfway point between field capacity and permanent wilting point, and this has been adopted as a working rule by a number of soil scientists and economists.

The Ruakura Institute in New Zealand estimate the growth of grass is affected when water supplies fall 40 mm below field capacity (25 per cent of volume) and ceases at a deficiency of 80 mm. Falcon and Gotsch (quoted above), however, estimate

Table 2.12. Experiments in over-watering.

	Total water input (mm)	*Marginal return (kg milled rice or dried lucerne/m³*)*
Rice		
Dry season: watering 5-daily	1650	0·011
watering continuously	4020	
Wet season: watering 5-daily	1580	0·024
watering continuously	3430	
Lucerne (Alfalfa)	620	1·94
	1000	1·57
Above	1250 1250	negative

Sources: India: Narayan Alyar, *Field crops of India* (Bangalore, 1944). USA: *USDA Yearbook*, 1955.

*Physical quantity of additional crop obtained from further watering divided by the number of cubic metres of additional water used. Three-day watering of rice was also tried, but gave results similar to 5-day watering.

for Pakistan that the 'half-capacity' rule would give only two-thirds of the maximum yield, but that a 'two-thirds capacity' rule would give 90 per cent. Such contentions, which have profound practical implications, underscore the need for further field testing of various hypotheses and weak theories which presently guide irrigation investment and operations.

Experiments on irrigated potatoes, grown in the cool climate of New Zealand, were conducted on the assumption that the farmer needs a single figure of soil moisture level as percentage of field capacity to tell him when to irrigate.[68] They were conducted at 20, 30, 40, and 50 per cent, indicating *optima* of 40 per cent for early and 30 per cent for later potatoes. Yield losses were significant if the farmer waited for lower proportions, but not serious if he began irrigating at higher levels of soil water content.

Hogg and others[69] measure moisture stress in sugar-cane growing in 'atmospheres' of pressure (1 atmosphere = about 930 cm of water). For comparison with Schofield's scale, described on page 00, wilting point at 15 atmospheres represents 4·15 on Schofield's scale; 20 per cent capacity is 2 atmospheres or 3·27; 40 per cent capacity is one atmosphere or 2·97; and field capacity 0·33 atmospheres or 2·5). Growth reaches its maximum at about 3 atmospheres (that is, below 20 per cent of capacity), and is not promoted by further watering—a surprising result.

Atmospheres water stress	6–5	4–3	3–2	2–1	1–0·5
Growth cms/day	0·6	0·9	1·2	1·2	1·2

They present a formula for annual yield as a function of water input:

Cane yield tons/ha/year = 307–7300 (water input in cms/per year)

Nowhere in the world is the maximum attainable: but some sugar-cane areas with annual rainfall of the order of 3000 mm can come close to it.

Flinn[70] constructs an interesting mathematical model in defence of the principle of irrigating when soil moisture falls to half capacity. He quotes experiments on maize[71] and Rhodes grass[72] to show that there is an irrecoverable loss of growth for each day on which transpiration actually falls below potential evapotranspiration. The extent to which this critical ratio depends on the level of soil moisture, Flinn points out, is still in dispute. Indeed it is. The situation could be described as chaotic. It is this confusion which B. S. Minhas, K. S. Parikh, and T. N. Sririvasan[73] set out to clear up by a brilliant piece of mathematical analysis. They point out that once we know the nature of the function relating actual/potential evapotranspiration to soil moisture content, we can give knowledge of initial soil moisture content and subsequent potential evapotranspiration and compute actual soil moisture content on any subsequent day.

The results of 15 experiments with soil moisture readings (confined to the top 120 cm of soil) were assembled by Minhas *et al.* Potential evapotranspiration was taken at 0·6 times pan evaporation, and immediate run-off of any excess rain-water was assumed (this hypothesis was tested). From these experiments, defined parameters were obtained for wheat growing in India. A further series of regressions on the experimental data were used to estimate the optimal distribution of the inputs of water between the periods 70–90 days from appearance (100 days in the case of some varieties), and the subsequent period to harvesting. Irrigation inputs from 0 to 250 mm were considered (Delhi has a rainfall of some 400 mm), with the following response:

Irrigation input (mm)	25·0	50·0	75·0	100·0	150·0	200·0	250·0
Marginal return (kg wheat/m^3)	20·50	11·30	6·50	4·20	2·00	0·72	0

The unirrigated yield in the area was 1·25 tons/ha. Although these deductions cannot be taken too literally, the rapidity in the falls of the marginal return is perhaps surprising.

Although Minhas and his collaborators have taken us considerably further with their elegant model, they have used only one functional form, and several configurations may be appropriate for a range of soil types and crops. The limitation to testing their model is at present a lack of good data. It is curious that plant physiologists largely ignore the economist's modelling of plant behaviour, mainly because they regard economic interpretations of plant growth as too simplistic, yet the main constraint upon further development and refinement of economic models is the deficiency of data that have to be provided by the plant physiologists.

Despite the millions of hectares under irrigation throughout the world, there is no definitive scientific understanding of the effects of alternative watering regimes upon useful crop production. Furthermore, the conceptual insights provided by the work of Beringer, Moore, Flinn, and others, discussed below on pp. 66–7, are still not normally incorporated in project design and operation. The almost complete lack of integrated research and development by crop scientists, economists, and operational irrigation engineers is perhaps a partial explanation of the poor performance of the majority of irrigation schemes.

It is clear that crop response to water supply is a complex of physical, biological, and biochemical processes.[74] Simple approaches, such as are utilized in taking total seasonal water inputs and relating this to yield, are extremely deficient. The impact upon yield of useful product of a particular degree of stress will depend upon the nature of the crop and the timing, severity, and duration of stress. Some crops are more resistant to stress than others—for example, sorghum is more resistant than maize. Nevertheless, it is necessary to consider response for all crops in a broader context than simple seasonal inputs.

Figure 2.4 illustrates the principle that the effect of stress is dependent upon the timing of stress. The solid curve *a* illustrates the normal growth curve where no stress occurs. Curve *b* illustrates a crop which suffers stress in the early stages of growth and never fully recovers. Curve *c* shows a crop that suffers stress at a critical period in the growth process (as peas do at flowering time). Curve *d* is occasionally found. For example, in the case of sugar-cane, temporary water shortage may be advantageous even if irrigation costs nothing. During the period of vigorous growth, sugar-cane will respond best, in terms of final yield of high quality juice, to soil moisture maintained at relatively low tensions. As harvest time approaches and during the ripening phase, sucrose accumulation is favoured by water stress.

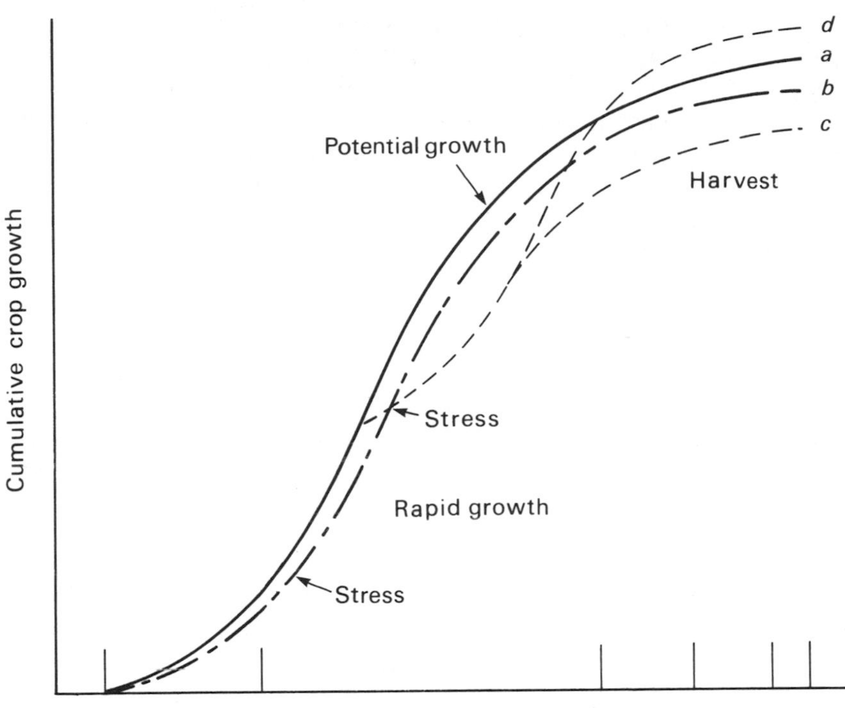

Figure 2.4. Hypothetical response to stress.

Economic models of crop response to irrigation

The Minhas team study (*op. cit.* note 73) is one of the most refined mathematical interpretations in the development of economic models of crop response to irrigation. Most of the models developed over the last 15 years can be placed in one of the four categories listed below with the names of some of the most notable contributors:

(1) crop yield as a function of total water inputs (early workers such as Dawson[75] and almost all current irrigation planners);

(2) yield as a function of an integrated soil moisture stress index (Beringer[76]);

(3) yield as a function of measures of moisture stress, which takes account of the timing and severity of the stress (Moore,[77] Yaron,[78] Minhas *et al.*[79]);

(4) yield as a function of a complex of variables, which takes account of the timing and severity of stress, crop condition, climatic conditions, stage of crop development, and so on (Flinn,[80] Palmer-Jones[81]).

The development, attributes, and limitations of the various contributions has been excellently reviewed by Irvin (1973)[82] and Palmer-Jones (1975). Despite these conceptual advances there is a persistence among agricultural planners in assuming that potential evapotranspiration is the optimum level of crop water requirement. This is no longer acceptable. However, more agronomic work is still required to find more precisely the economic watering level for the major irrigated crops.

Crop response to irrigation using fertilizer

Lack of water, which lowers yield, may be mitigated by the application of chemical fertilizers; the widely held view that the effects of drought are accentuated by such addition is erroneous. The issue was dealt with magisterially by the US President's Science Advisory Committee.[83]

> It is often thought that the use of fertilizers requires a parallel increase in water applied. Unless the application of fertilizer results in dense foliage which provides a much more complete coverage of the soil surface and hence allows a higher rate of evapotranspiration/unit area, no significant increase in water use is to be expected with increased fertilizer. In may cases, vegetation cover is already sufficient and fertilizers provide an important means of increasing crop production per unit of water supplied.

This latter proposition was directly confirmed by Penman,[84] who surveyed the results of experiments on three strains of grass at Hurley and Kew:

Nitrogen input kg/ha	0	19	38	76
Grass dry weight yield kg/m³ transpired	1·8	1·9–2·0	2·4	2·8–3·0

Another important study[85] on fertilized grass (see table 2.13) showed that fertilizer, far from increasing the plant's demand for water can, in effect, serve as a 'substitute' for it. In this study, in view of the variability of the rainfall (the work was done in the UK in 1953, a fairly normal season with 338 mm of rain from April to September), no predetermined amounts of water were used, but in the various experiments the rule was made to begin watering when the water deficits in the soil had reached 17, 34, and 51 mm respectively. In the best-fertilized plots the grass did not seem to be much affected if watering was delayed till the

51-mm deficit, and the gain over the unirrigated land was much less than the unfertilized or less-fertilized grass.

Table 2.13. Combined response to fertilizers and irrigation (grass dry weight tons/ha/yr).

	Unfertilized	*Fertilized with 314 kg/ha nitro-chalk*	*Fertilized with 628 kg/ha nitro-chalk*
Unirrigated	8·5	11·2	14·0
Irrigated at			
51–mm deficit	9·7	12·2	15·2
34–mm deficit	10·3	13·1	15·3
17–mm deficit	10·8	13·2	15·6

An even more thorough series of experiments was carried out in Pakistan.[86] The nitrogen fertilizer inputs were all accompanied by 67kg/ha P_2O_5. Water inputs represent additions to the pre-sowing irrigation of 10 cm, as shown in table 2.14.

Table 2.14. Wheat yields with various levels of water and fertilizer inputs.

		Yield (tons/ha) Nitrogen input (kg/ha)					
		0	*34*	*67*	*111*	*136*	*168*
Traditional Pakistan Wheat Type C273							
Water input (cm)	23·0	1·1	1·6	1·6	1·6	1·8	2·6
	30·0	0·8	1·2	1·6	2·0	2·3	2·2
	38·0	1·0	2·0	1·8	1·9	2·5	2·2
	46·0	1·4	1·6	2·3	2·4	2·2	2·6
New strain Mexipak Wheat							
Water input (cm)	23·0	1·1	1·5	2·3	1·7	2·2	2·6
	30·0	1·1	1·4	2·1	2·3	3·0	2·7
	38·0	1·2	2·0	2·1	2·7	3·4	3·0
	46·0	1·2	1·7	2·2	3·2	3·3	4·0

What appear to be occasional negative responses to increased inputs of either water or fertilizer are perhaps due to aberrations in particular experiments. It seems clear that nitrogen inputs increase responsiveness to water in this case, and that the new strain can attain its maximum potentialities only with high inputs of both water and fertilizer.

Similar results[87] were obtained with the growing of wheat in

Table 2.15. Wheat yields and water inputs.

Water inputs (mm)	Yield (tons/ha) Nitrogen input (kg/ha) 0	50	100	150
440	0·34	1·02	1·77	2·19
550	0·28	1·54	2·73	3·80
740	0·43	—	3·10	4·41
780	0·54	1·85	3·47	4·02

Mexico during the dry winter November–May (see table 2.15 below). Six years' results on experiments with barley[88] in England are shown in table 2.16. The more heavily fertilized crops show less response to additional water. There is also general agreement that phosphatic fertilizers increase drought resistance, probably be accelerating their maturity. However, nitrogenous fertilizers may be dangerous in some climates if they delay the plant's maturity beyond its normal growing season into a dry season; this may be the origin of the fear that they lead to a general increase in water needs.

Table 2.16. Fertilized and irrigated barley.

	Fertilized with 25 kg/ha nitrogen (element)	Fertilized with 50 kg/ha
Tons/ha yield; Unirrigated	3·46	3·95
Irrigated	3·83	4·08

This response to nitrogen under irrigated conditions is not universally found, as is shown by the maize experiments in Israel, reported by Shalhevet *et al.* (*op. cit.*, note 50). This Israeli work indicates that the interaction between the irrigation regimes and nitrogen fertilizer is of the limiting factor type, and is shown in fig. 2.5. Increases in water applications over the range studied all show a positive marginal response to water. However, it is at the highest levels of water application that the marginal return to water is highest, with approximately 1 kg/m^3 with high rates of fertilizer, but close to zero with no fertilizer. Over the range of applications from 600 to 850 mm, the highest marginal return to water comes from the highest applications of fertilizer, indicating that the maximum yield per hectare using a combination of water and fertilizer had not been reached in the experimental designs

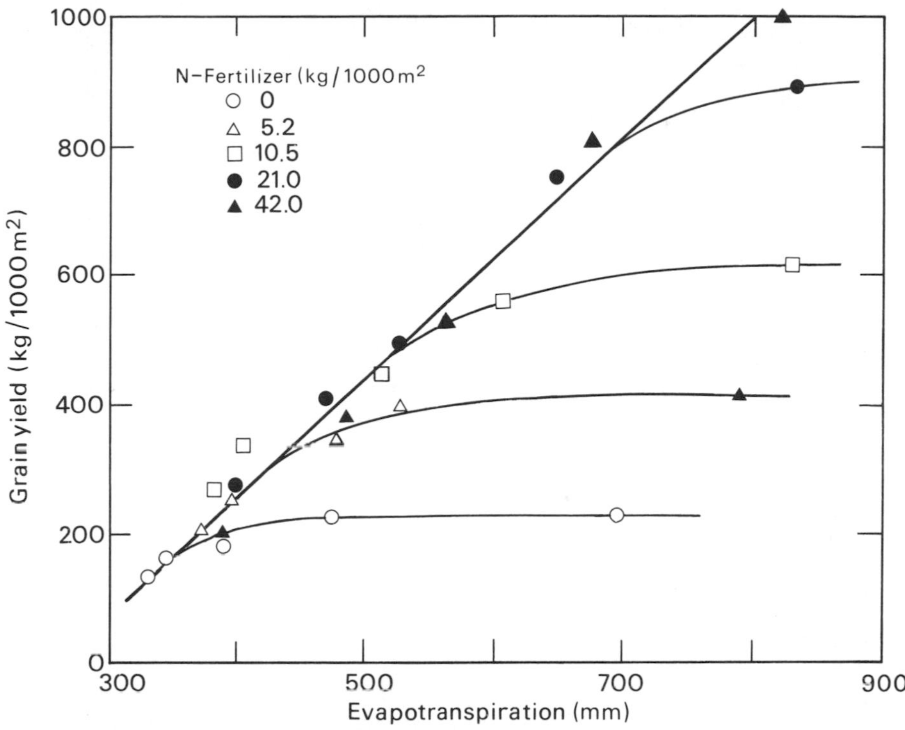

Figure 2.5. Relation between seasonal evapotranspiration and yield of corn grain at five levels of N-fertilization in Israel.

Source: Shalhevet *et al.*, *op. cit.*, note 50, p. 252.

adopted to date. At higher applications of both factors a higher yield per unit of land would be possible. However, the slope of the water response functions alone cannot help in determining optimum irrigation policy. More information is required of the value of output and the costs of complementary resources required for crop production.

William, Stiles, and Turner[89] conducted a series of experiments to test different routines for grass irrigation. The experiments varied a little between years, but in general it appeared that the routine yielding the best results was to water crops whenever the deficit reached 40–50 mm, applying 25–50 mm of water. In all the experiments, superphosphate was applied at the rate of 94 kg/ha and the same quantity of 60 per cent muriate of potash. The nitrogen treatments varied. The quantities of water applied as a result following this routine varied greatly among years, from 50 mm in the very wet summer of 1958 to 330 mm in the exceptionally dry summer of 1959. In this dry year, the gain of dry

Table 2.17. Rice yields related to water control measures and the use of additional inputs.

Degree of water control	Use of additional inputs		Rice yield* (tons/ha) 1971	1972	1973	1974	Rice yield obtainable as average (tons/ha)
No water control (rain-fed, uncontrolled flooding)	nil	Laos	1·2	1·2	1·3	1·3	1·3
Successive introduction of water control							
(a) elimination of floods	nil	Democratic Kampuchea	1·5	1·4	1·3	0·9	1·5
(b) elimination of drought	low fertilizer application	Burma	1·7	1·6	1·7	1·7	
		India	1·7	1·6	1·7	1·6	
		Thailand	1·9	1·8	2·0	1·7	2·0
(c) improved water control (irrigation and drainage)	low to medium fertilizer application	Pakistan	2·3	2·4	2·4	2·1	
		Vietnam Republic	2·4	2·4	2·5	2·5	
		Sri Lanka	2·4	2·4	2·3	2·9	
		Malaysia	2·5	2·4	2·5	3·0	
(d) sophisticated management practices (mid-season drying)	high fertilizer use + improved seeds and pest control + diversification	Republic of Korea	4·6	4·6	4·9	4·8	5·0
	mechanization	Japan	5·2	5·8	5·9	5·8	6·0
Experimental conditions							10·0

*All figures relate to paddy rice (1 ton paddy approximately = $\frac{111}{322}$ ton rice).

Source of yield figures: *The State of Food and Agriculture*, 1971, 1972, 1973, and 1974 (FAO, Rome) from FAO, *Water for Agriculture*, Document to UN Water Conference, Mar del Plata, Argentina, 1977.

weight yield of grass on unfertilized land was 1·35 kg/m^3 of water applied; on land fertilized with nitro-chalk at the rate of 500 kg/ha the gain was 1·83 kg/m^3. In the wet year of 1958 neither pasture showed an appreciable gain. In the moderate rainfall years of 1956 and 1957, on the other hand, the unfertilized land showed a greater gain per unit of water than the fertilized.

In a careful study, Will[90] estimates that irrigation raises what he calls the 'degree of exploitation' of fertilizer from 55 per cent to 63 per cent. It raises requirements of potassium and phosphorus relative to requirements of nitrogen. While increasing stem and leaf formation, irrigation reduces root growth. Plants put more effort into root growth when water is scarce. This can lead to moisture stress problems if irrigation is removed—for example, by an unreliable supply, which may help to explain farmers' reluctance to adopt irrigation when delivery is erratic. The FAO has recently produced evidence of the increased rice yields that occur under Asian field conditions from joint application of increased water control and additional inputs such as fertilizer. This is reproduced as table 2.17.

It is clear that irrigation when applied jointly with other inputs such as fertilizer, crop protection, and genetically superior varieties produces a yield increase which, in total, is greater than the increase which would occur if each input was applied separately. Thus, if irrigation, fertilizer, crop protection, and high yielding varieties each increased yield by 10 per cent, the total yield increase if they were applied jointly would be more than 40 per cent, say 70 per cent. This more than additive or complementary response to irrigation explains the World Bank contention that in India, with existing technology, yields could be increased by improved agronomy between 10–30 per cent on rain-fed land but by 25–50 per cent on irrigated land.[91]

The problems of interpreting field experimental data when joint or interactive response exists is discussed by S. K. De Datta *et al.* (1978) (*op. cit.*). They illustrate the type of insight obtainable by data relating to experiments conducted on a farmer's paddy field in Laguna, Philippines. The table on p. 66 shows that if either insect control or fertilizer is applied alone no significant yield increase is obtained but when they are jointly applied yield increases of 1t/ha are obtained. Further examples of this kind are provided by Gomez (1977).[92]

In village studies conducted by the International Rice Research Institute in 36 villages of six countries of South and South-east Asia it was found that adoption of modern rice varieties varied considerably from village to village. In some areas 90 per cent of

Table 2.18. Factors contributing to rice yield gap in the Philippines

	t/ha
Actual farm yield	3·69
Potential farm yield	4·95
Yield gap	1·26
Individual contributions:	
Insect control	0·07
Fertilizer	0·11
Weed control	−0·29
Joint contributions:	
Insect control and fertilizer	0·99
Insect control and weed control	−0·03
Fertilizer and weed control	0·28

fields were sown but in other areas less than 20 per cent of the areas was under modern varieties.[93] It was found that where irrigation provided a high degree of water control the rate of adoption was higher (other factors were the variety, seasonal suitability and its profitability). The first generation of modern varieties were dwarf and although they had the highest yield potential they were vulnerable to fluctuating water levels. They have proved popular in irrigated areas with high levels of solar radiation.

Project planning and crop response to irrigation

Establishing the level of response to irrigation is a key issue in planning. It is, however, an extremely complex procedure. In this chapter it has been shown that crop response to irrigation depends on a number of other factors, such as growth in previous time periods, addition of fertilizer or other inputs which interact with water, and the effects at various stages of growth of the climate, pests, and diseases.[94]

Normally, project planners approach their task by calculating the supply of that additional quantity of water which gives maximum crop yield per hectare. This usually involves matching the supply of water to the potential evapotranspiration. A progressive approach, however, is the analysis of the problem from a different point of view: that of selecting a simple response or production function (which indicates the returns to additional water as shown in curve (*i*) of fig. 2.1) and determining from this the best operating conditions. Unless there is virtually free water the most economic application will be less than that required for maximum evapotranspiration and less than that for maximum yield per hectare.

Difficulties in observing this procedure often stem from local conditions. There are all the usual problems with production function analysis including specification of the functional form (linear, quadratic, and so forth), estimation or establishment of the empirical measures, and interpretation of the statistical relationships.

Specification is usually achieved using a quadratic or sigmoid functional form, and this forces certain mathematical properties upon the relationship which may not be appropriate.

Estimate is difficult because data may not be available, or unavailable in the form required. Knowledge may be limited; for example, sensitive periods in the growth cycle for water stress may not be known, or there may be uncertainty about the suitability of canal water or slightly saline tubewell water for the crops to be cultivated. Data which may be available from experimental farms may not be applicable to the planning of small peasant farms with a very different endowment of resources. Figure 2.2, on page 50, demonstrates from data from Pakistan that considerable differences exist between experimental and practical farm situations.

The problem of establishing appropriate levels of water input for project planning and agricultural extension should be approached in the following way:

(1) Investigate farm level yields and their determinants by a farm survey.

(2) Use these data to identify the major determinants of yields and irrigation practices. Farmers own opinions, solicited probably after participant observation, would be useful for this.

(3) Identify the institutional social, political and economic constraints on actual water productivity.

(4) Design, execute, and analyse suitable irrigation experiments which separate time and quantity (and frequency) of irrigation effects. The interaction of weather and response to irrigation must also be taken into account by (a) yield-weather analysis; (b) analyses of irrigation experiments in different years and ecologies.

Further difficulties arise for the project planner in estimating crop response to irrigation when, as on most farms, several products are envisaged. The optimum water input for one crop may be totally unsuitable for others on the farm. Furthermore, in a multiproduct enterprise such as a farm, with a variety of resource constraints, optimising water application to one crop is no guarantee that it will be optimum from the whole farm viewpoint. These problems will be treated in detail in chapter 8.

CHAPTER 3 WATER RESOURCES AND THEIR EXPLOITATION

WORLD WATER RESOURCES

Global water resources may be summarized as 97 per cent, or 1350×10^6 km^3 (10^9 is the multiplication factor for m^3 to km^3), in the ocean and 2 per cent, or 26×10^6 km^3, in ice-caps and glaciers (enough to raise the sea level appreciably if they melted). The balance of 1 per cent is made up of groundwater (7×10^6 km^3), saline and freshwater lakes ($0{\cdot}26 \times 10^6$ km^3), soil moisture ($0{\cdot}15 \times 10^6$ km^3), plus negligible amounts in rivers and biological systems. Movement of water from one state or location to another occurs continuously. Annual evaporation is 516×10^3 km^3 per year, 86 per cent of it from the ocean. Precipitation of this moisture also falls largely on the seas (80 per cent).[1] From these estimates it is clear that only a small proportion of the world's water is likely to be available for agriculture. Ninety-nine per cent is either saline or frozen. Of the rainfall on the land, 28 per cent ($29{\cdot}5 \times 10^3$ km^3) reaches the sea, the remainder being evaporated from the soil, vegetation, or water surfaces. Even so, the amount reaching the sea is an annual average of about 6500 m^3 per head of the world's present population.

These statistics are very approximate, because most countries have widely variable estimates of water availability. For India, where statistics have been kept for decades, estimates of surface water flows range from 144 million hectare metres per year to 188 million hectare metres per year. The current estimate of mean annual surface flow used for planning purposes by the Central Water and Power Commission is 167 million hectare metres. It is assumed that the 'dependable' flow is 75 per cent of this, or 125 million hectare metres; that utilizable flow is 66 million hectare metres, which, together with groundwater, results in a total availability of 87 hectare metres per year (Irrigation Commission, 1972). However, other reasonable assumptions on availability

Notes and references for this chapter begin on p. 254.

and variability of flow give a broad range of estimated availability.

Run-off estimates

The proportion of precipitation which reaches the sea varies widely throughout the world. Some 60 per cent of global run-off occurs in Asia and South America. Where water is needed most in hot, arid climates, the proportion of the rainfall which evaporates is highest and the proportion which reaches rivers is lowest. In India, even in the highly humid Brahmaputra catchment, the run-off is only 60 per cent of inputs (see table 3.1).

Table 3.1. Potential utilization of Indian rivers.

	Area (million ha)	*Average rainfall (m)*	*Average run-off (m)*	*Stream flow* (10^{10} m^3)	*1950 use for irrigation* (10^{10} m^3)	*Proposed use of irrigation and power* (10^{10} m^3)
West coast river basins, excluding Indus	49·2	1·22	0·63	31·0	1·4	5·1
Indus basin, excluding W. Pakistan	35·4	0·56	0·22	8·0	1·3	3·0
East coast river basins, excluding Ganges and Brahmaputra	119·0	1·09	0·34	41·2	3·0	16·1
Ganges basin, excluding E. Pakistan	97·7	1·11	0·50	49·0	3·2	6·1
Brahmaputra basin, excluding E. Pakistan	50·6	1·22	0·65	38·2	0·4	4·4
Rajputana (draining inland)	17·0	0·29	0	0	0	0
All India	365·9	1·06	0·45	167·4	9·3	34·7

Source: Indian Water and Irrigation Commission

Geographers have tried out a large number of formulae to express run-off as a function of rainfall under different circumstances. Application to one climate of a formula designed for another may lead to ludicrous results, as happened when a prominent Australian engineer used a German formula to estimate run-offs, and then proceeded to prepare extremely ambitious plans for exploiting non-existent water. A formula for low rainfall countries, developed in Tunis by Tixeront,[2] is as follows:

$$\text{Run-off in metres} = A\ (\text{rainfall in metres})^3$$

where, for Tunis, A lies between 0·25 and 0·4. Data for Australian[3] rivers also indicate a cubic relationship with a coefficient of 0·2. This covered a range of run-off proportions from about 0·05 in some West Australian rivers to a maximum, in the very high rainfall tropical watershed of the Tully in North Queensland, of 0·62.

Roederer's data for Morocco[4] also show something like a cubic relationship, with the constant at 0·22 for the Atlantic coastal rivers, 0·35 for rivers of the Rif, and about unity for the Saharan rivers of the Atlantic–Atlas slope. For the latter, however, a better fit is obtained with an exponent of 2 and a coefficient of about 0·7.

Data for Sri Lanka[5] appear to indicate a value of A of 0·15–0·2, and an exponent of 2 instead of 3. The fact that the high lands, unlike those of India or Tunis, are covered with dense, deep-rooting jungle, may increase transpiration and reduce run-off. On the other hand, the increased evapotranspiration due to tree cover must not be exaggerated. Measurements in the Harz Mountains[6] showed evapotranspiration of 0·58 m/yr by trees and 0·52 by grass nearby (out of annual rainfalls of 1·25 m and 1·22 m respectively).

However, in hotter climates[7] evaporation from forested areas may be expected to be substantially larger than from grass because (1) deep-rooted trees continue evaporating, whereas grass is forced to close stomata and cease growth or to die back in dry weather; (2) considerable rainfall is intercepted on the leaves and bark of the trees, and is evaporated by wind and sun without serving to cool the plant tissues.

In California it has been observed that pine forests evaporate more from a given area than do eucalyptus, eucalyptus more than deciduous trees, deciduous trees more than grass. These considerations often lead water supply engineers, to whom each extra centimetre of water is valuable, to call for the clearing of forests on watersheds. This may bring dangers of soil erosion, and siltation of reservoirs unless cover on the catchment is quickly re-established and grazing is subsequently regulated.[8]

Average run-off in hilly country in Italy, whether grassed or forested, is estimated[9] at 20–25 per cent. For the 150,000 km^2 of England and Wales, with an average rainfall of 905 mm, run-off is estimated[10] at 50 per cent, or $6{\cdot}8 \times 10^{10}$ m^3/yr. Central Asia is generally regarded as arid; but it is claimed that the total flow of streams in Sinkiang is no less than $10{\cdot}7 \times 10^{10}$ m^3 *per annum*.[11] Taiwan has a total rainfall of $8{\cdot}88 \times 10^{10}$ m^3/yr of which 4·04 runs off—a highly favourable proportion for a sub-tropical climate. Taiwan's exceptionally dense population, living mainly by rice growing, still only captures 27 per cent of the run-off.

The Indian estimates of the proportion of run-off which could be captured (including that used for hydro-electric generation) amounted to 40 per cent for the East Coast rivers other than the Ganges and Brahmaputra, but only to 20 per cent for India as a whole. In the more humid Ganges and Brahmaputra Basin it is not proposed, in the present system of river plans, to attempt to conserve more than 20 per cent of the flow. The Brahmaputra conservation will be entirely for hydro-electric power—in that district any water required for agriculture can be readily obtained from tubewells, wells, or ponds. Both basins suffer from a lack of potential dam sites.

Under Australian conditions,[12] where rainfall is irregular and the ratio of stream run-off to rainfall low, it is estimated that 20 per cent of the flow of a river is generally the maximum utilizable proportion, and that the present irrigated crop area could be extended at most two- or three-fold, and even then only at considerable cost. Aird[13] points out that the volume of water which has to be stored in Australia, per unit of land irrigated, is very large, sufficient to cover the whole irrigated land to an average depth of 1·4 m, as against 1·07 m in the United States, 27 cm in Egypt, and only 9 cm in India.

Dependable flow

Data of the availability of river water represent in average terms a stochastic relationship between water flow and time. The variation around the mean in various rivers in a defined season will differ widely, depending upon many factors, including the reliability of rainfall and the proportion of run-off from snow melt. Rivers such as the main branch of the Indus, which rely largely upon snow melt for discharge, will have a narrower range of variation than, for example, tributary rivers such as the Jhelum, which is dependent upon monsoon rainfall. Some rainfall patterns are more reliable than others, but sunshine hours and temperature levels are more predictable than precipitation.

Given a wide variation, it is odd that certain conventions or 'house rules' for availability such as 'use the 1 in 5 dry-year flow' or the '1 in 10 dry-year flow' are adopted by engineering consultants and irrigation departments for widely variable locations. At the beginning of this chapter it was noted that, in Indian irrigation projects, water availability is assessed using 'dependable flow', which is 75 per cent of mean flow. In Pakistan, the Indus basin storage works were designed using mean flows because it was assumed that shortfalls could be made up using groundwater.[14]

In the early years of irrigation development, where river flow data was scarce and information on agricultural water use and the impact of drought even more sparse, conservative assumptions such as design for 1 in 5 dry-year flows were perhaps understandable. However, the inclusion of economists in planning teams should encourage the use of marginal analysis and partial budgets to assess as objectively as possible the costs and benefits of various levels of installed capacity. High levels of installed capacity (or large-scale projects) will increase the risk from shortfall in water supply at various times of the agricultural year. The impact of a given level of shortfall is location specific and the acceptability of this risk depends, amongst other things, upon the scale of the project in relation to the local economy, the existence or not of other related risks, and the income status of the persons at risk.

In some agricultural economies it may be impossible to sustain the losses associated with a 1-in-20 year drought in a critical month, whereas in other countries, or in other months, the 1-in-2 year shortfall might be acceptable. It is sufficient to note that in planning projects, particularly once water becomes a scarce factor, estimating the effect of marginal adjustments in water supply can be a very important component of the economist's work and often crucial for determining the scale of the project.

Economists have carried out numerous studies outlining the concept of an economic loss function.[15] This measures the economic effect on a system of a shortfall in water supply that leads to a failure to achieve planned output. System simulation can incorporate a loss function. It is assumed that data is available (or obtainable) to indicate whether, in the event of a water shortfall, it is rational to reduce per ha applications of water or reduce the area cultivated. Practical matters relating to local circumstances and custom are also important. It may be that any shortfall is shared evenly by all farmers (who may water all fields inadequately or concentrate water resources on fewer fields); it may be that farmers nearest the source satisfy their needs first, it may be that historical precedent is important and the first farmers to use irrigation in the region will receive as priority full supply and others must await any surplus. In the more feudal parts of Asia allocation appears to be on the basis that 'might is right'. Empirical measures of behaviour and assessment of its efficiency from an economic or social point of view are worthwhile investigating before designing a system.[16]

River flows and abstractions

The total flow of the Indus is 20·7 × 10^{10} m^3/yr (9 per cent originating in Afghanistan, 14 per cent in Tibet, 17 per cent in Pakistan, and the rest in India). The extreme variability of the flow is demonstrated by the figures shown in table 3.2 (1921–51

Table 3.2. Indus flows (10^{10} m^3/month).

	Minimum monthly flow		*Maximum monthly flow*			*Lower Indus*	
	All rivers	*Indus*	*All rivers*	*Indus*	*Average flow*	*Used in Punjab*	*Used in Sind*
January	0·3	0·2	0·7	0·3	0·42	0·18	0·14
February	0·3	0·2	0·8	0·3	0·43	0·21	0·15
March	0·4	0·2	1·4	0·3	0·53	0·27	0·11
April	0·6	0·3	1·7	0·8	0·77	0·28	0·12
May	0·8	0·4	3·3	1·8	1·40	0·46	0·20
June	1·7	1·1	4·5	2·7	2·47	0·66	0·41
July	2·8	1·7	6·9	4·1	3·30	0·70	0·64
August	2·6	1·5	6·7	3·4	4·01	0·90	0·76
September	1·2	0·7	3·5	1·4	2·49	0·67	0·44
October	0·5	0·3	1·1	0·5	0·98	0·41	0·23
November	0·3	0·2	0·6	0·3	0·57	0·24	0·17
December	0·3	0·2	0·6	0·3	0·42	0·18	0·09

Source: Pakistan Irrigation Authorities.

average) of stream-flow entering the Pakistan province of Punjab, and 1932–40 data of utilization of lower Indus waters in Punjab, and the down-stream province of Sind. In considering these latter figures, it must be borne in mind that lower down the stream, the Indus will have absorbed other rivers as tributaries and also to some degree regularized its flow. It is worth noting that the degree of control over the Indus is such that for several months of the year, no water now reaches the sea.

The World Bank skilfully negotiated an agreement between the ill-disposed countries of India and Pakistan in 1960 on division of the Indus waters. The Bank now estimates the water due to reach Pakistan, after full implementation of the Indus

Water Treaty, which has provision for large volumes of surface water storage, at $17{\cdot}9 \times 10^{10}$ m^3/yr, as against canal diversions in the 1960s of only $9{\cdot}8 \times 10^{10}$ m^3/yr.[17] India has the right to withhold virtually all flows on the Indus tributaries Ravi, Beas, and Sutlej, which is about $3{\cdot}1 \times 10^{10}$ m^3/yr. Although this agreement helped to avert war (for a period) between these countries, it also resulted in the construction of two expensive short-life, on-stream storage areas in Pakistan (Mangla and Tarbela) and the extension of irrigation in India of some largely unsuitable sandy and saline areas.

Though more famous, the Nile has much less annual flow than the Indus—only $8{\cdot}3 \times 10^{10}$ m^3/yr. For thousands of years Egypt has exploited the natural summer rise of the river and its flat riverside lands to practise a simple form of flood irrigation. This was augmented by storage at the Aswan Dam,[18] beginning at $0{\cdot}10 \times 10^{10}$ m^3 in 1902, 0·24 in 1912, 0·5 in 1933, and 0·8 in 1937, with the construction of the Jebel Aulia dam. By the 1950s total abstraction amounted to $4{\cdot}8 \times 10^{10}$ m^3 in Egypt, and $0{\cdot}4 \times 10^{10}$ m^3 in the Sudan, totalling 63 per cent of the entire river flow. But the ambitious new Aswan High Dam, whose financing led to a world political crisis in 1956, has been planned to capture the entire stream flow. The financial cost will be almost entirely offset by the value of the hydro-electric power generated. Non-financial costs include an estimated $1{\cdot}0 \times 10^{10}$ m^3/yr evaporation from the huge new lake to be created, and salination and disturbance of the fisheries at the mouth of the river. The final distribution of the new net flow ($8{\cdot}3 \times 10^{10}$ m^3 less $1{\cdot}0 \times 10^{10}$ m^3 evaporation) is to be $6{\cdot}2 \times 10^{10}$ to Egypt and $1{\cdot}1 \times 10^{10}$ to the Sudan.

The old, but not yet re-negotiated, Egypt–Sudan Nile Waters Agreement provided that during the dry period, 1 January to 15 July, with a total flow of only $1{\cdot}54 \times 10^{10}$ m^3, the Sudan could take only 0·15. During the flush season, the Sudan could take $1{\cdot}4 \times 10^{7}$ m^3 daily into the Gezira canals, basin-irrigate certain lands, and fill up storages, up to a total right of $0{\cdot}8 \times 10^{10}$ m^3—half of which remained unused. This agreement will have to be re-negotiated when the new demands of riparian states are formulated (including Uganda, Ethiopia, and Kenya). Plans to by-pass the Sudd, where much base flow is evaporated, are, though found to be technically feasible, likely to be shelved for environmental reasons. We are now in an era when even an apparently obvious project such as draining a water-wasteful malarial swamp has to be carefully evaluated because of potentially harmful environmental impact.

Water abstracted for irrigation, it is sometimes said, is only a moderate proportion of total abstraction, when we consider the demands for power plants, industry, domestic use, and so on. But this is to confuse gross and net demand. The large quantities of water circulated through thermal power stations are returned almost completely to the rivers, albeit somewhat warmed; and industrial and domestic water is also fairly completely returned, subject to various degrees of contamination (which can be removed, at a cost, by sewage treatment). If, as is usually the case, we grudge the expense of sewage treatment, and water is fairly abundant, we 'demand' large stream flows to 'dilute' our sewage to save us the expense of treating it.[19]

If, however, we were willing to spend more on the treatment of effluents, the volume of water needed to dilute them to a tolerable level would be less. The vast cost of cleaning polluted rivers after the event would be avoided. The time may come when we will decide that we can no longer afford to use a thousand tons of water to carry away one ton of waste. This will arise when the value in alternative use of a thousand tons of fresh water is less than the treatment costs of one ton of waste.

Intrusion of saline water to estuaries

Increasing levels of abstraction and storage for irrigation and other use is resulting in low levels of zero flow in many estuaries for all or part of the year. This is causing encroachment of saline water into previously freshwater estuaries, which can be very detrimental to cultivable land, irrigated or not, in the estuary region. Similar problems exist where public water supplies are abstracted from what was originally a freshwater source.

The Hydraulics Research Station in the United Kingdom has been working on computer methods to predict the long-term effect of such salination using minimal data. Urgency was injected into this situation by the recent Sahelian drought, which had disastrous effects upon water qualities of some West African rivers used for irrigation hundreds of kilometres inland.[20] Prediction would be valuable, but prevention is more important. A number of countries are investigating barrages and other devices to prevent the encroachment into fresh-water resources of saline water. The UK Hydraulics Research reports, in the same source, research into inflatable dams for irrigation schemes made of flexible but inextensible material. These are essentially water-filled plastic sausages, designed to be used at sites where rigid structures would be too costly or take too long to construct.

However, the most important use might be in estuaries to prevent the intrusion of saline water.

WATER REQUIREMENTS FOR AGRICULTURE

Measured net, the requirements of agriculture are seen to be very large in relation to all other uses. For example, 10,000 m^3 of water will be sufficient to:

(1) irrigate 0·5 hectares of rice, 0·8 hectares of cotton, 1·2 hectares of wheat;

(2) supply water to 100 nomads and 450 head of stock for 3 years;

(3) supply water to 100 farming people through standpipes for 14 years;

(4) supply water to 100 farming people through house connections for 4 years;

(5) supply 100 consumers in a modern industrial city for 2 years;

(6) supply 100 luxury hotel guests for 55 days.

Desert areas are ideal for agriculture if water is available. However, they are also suitable for modern industrial development.

> The water available in the area would produce great wealth if devoted to the comforts of those who come to enjoy the climate of Arizona and New Mexico, mild winter, luminosity, dryness; to industries exploiting local mineral resources, and also highly developed industries requiring little raw material, e.g. electronics, seeking to attract personnel who wish to live in an agreeable climate. Using large quantities of irrigation water to produce cereals and fodder crops in these deserts appears wasteful in comparison. The crops obtained cannot pay the true price of the water and cannot face the competition of the produce of unirrigated land.[21]

In 1960 the Select Committee on National Water Resources of the US Senate[22] made an estimate and projection of US water requirements (see table 3.3). In the face of these expected demands, rising to an annual level of $96{\cdot}6 \times 10^{10}$ m^3 units by the end of the century, the average flow of all streams in the United States was estimated at 187×10^{10} m^3, nearly twice the end-century demand. However, the stream flow is very irregular between seasons, and large costs have to be incurred for storage if the total flow, or indeed any large fraction of it, is to be used.

Table 3.3. Annual water requirements in the United States (10^{10} m^3)

	1954		*1980*		*2000*	
	Gross	*Net*[1]	*Gross*	*Net*[1]	*Gross*	*Net*[1]
Agriculture	24·2	14·3	23·0	14·4	25·4	17·4
Power stations (cooling)	10·2	0·1	35·6	0·2	59·1	0·4
Manufacture & mining	4·6	0·4	14·4	1·3	32·0	3·0
Domestic & other	2·3	0·3	3·3	3·9	0·5	5·8
Increased requirements for lakes		0·8		9·8		13·4
Required for dilution of effluents		70·7		46·0		61·6
Required minimum drawings upon stream flow (sum of above)		85·8		72·2		96·6
Additional requirements to be provided if possible:						
Hydro-electric		51·9		85·3		88·0
Navigation		38·8		32·9		30·5
Fishing		10·8		23·6		33·3
Population (millions)		162		244		329

[1] After allowing for water returned to streams.

If enough water can be stored, on the one hand, and economized by better treatment of waste, or reduction of excessive irrigation demand, on the other hand, there are further claimants for large quantities of water, in hydro-electric generation, navigation, and the improvement of fishing.

Available water supply

In the western United States it is estimated[23] that water supply available for irrigation averages 40×10^{10} m^3 annually, of which 9·6 are at present used to irrigate 10 million hectares (that is, average supply, before losses in transit, about 1 metre per hectare). These ideas seem to conflict with the official estimate. The maximum use is put at 16×10^{10} m^3 to irrigate 17 million hectares.

Japan[24] in 1970 was still only using 16 per cent of run-off. The rate of utilization is expected to rise to 26 per cent by 1985, even though industry is expected to re-cycle 70 per cent of its (immensely increased) demand. Local shortages are expected in the Tokyo and Osaka regions—this was one of Prime Minister Tanaka's reasons for proposing decentralization. Irrigation demand is expected to be stabilized. Actual and estimated water use is presented in table 3.4 below.

Table 3.4. Water use in Japan.

	Water use (10^{10} m^3/yr)			
	1970 actual		*1985 estimated*	
	Gross	*Net*	*Gross*	*Net*
Agriculture		5·34		5·34
Industry	3·6	1·8	32·0	5·7
Domestic		0·92		2·14
		8·06		13·18

Despite the widespread public outcry at water restrictions during the 1975 and 1976 drought, the total run-off in England and Wales (6·8 × 10^{10} m^3/yr) is not heavily drawn upon,[25] as can be seen in table 3.5.

Table 3.5. Water use in England and Wales.

	(1967 10^{10} m^3/yr)	
	Licensed	*Abstracted*
Electricity generation	2·30	1·15
Industry	0·75	0·62
Public water supply	0·76	0·44
Agriculture	0·05	0·03

Of the public water supply, only about one-third is metered, mostly for industrial use. Domestic supply, of which virtually none is metered, amounts to 0·15 m^3/person/day, compared with 0·12 in France and Germany, and about 0·4 for the United States. Total domestic demand for West Germany was 0·22 × 10^{10} m^3 in 1957, in comparison with an irrigation demand of 0·20 × 10^{10} m^3.[26]

For Israel much smaller resources are available. It is possible that the Israeli experience gives a vision of what could be, and may be, achieved in future years in other arid countries. It is estimated that the annual water potential is 1·7 × 10^9 m^3 per year derived from the sources shown in table 3.6. Virtually all the non-saline water is now used, 77 per cent of it in agriculture. An area of 186,000 ha is irrigated annually, some 43 per cent of total area under cultivation. Average water use is 6,000 m^3/ha compared to 8000 m^3/ha some twenty years ago. During the period 1964–74, while the annual amount of water used for agriculture remained relatively stable, the production of citrus increased by 200 per cent, cotton by 300 per cent, and fruit and vegetables by somewhat lower amounts. The same amount of water produced in 1975 two and a quarter-times the agricultural output of 1959. Some 87 per cent of the irrigated area is sprinkler-irrigated, 10 per cent is drip irrigated, and only 3 per cent is irrigated by gravity channels. A gradual shift to such methods and increased levels of efficiency can be expected in the next decade in the oil-rich countries and the temperate industrialized countries. However, in most of Africa and Asia the abundant labour supply and capital shortage makes development of capital intensive irrigation an inappropriate means for obtaining efficient utilization of water resources.

Table 3.6. Estimated water resources of Israel (10^6 m^3).

Surface water, mainly in the Jordan	425	(25%)
Groundwater and springs	1020	(60%)
Brackish water	170	(10%)
Flood water	85	(5%)
Total	1700	(100%)

Source: International Irrigation Information Centre. *Irrinews,* **13** (Volcani Centre, Bet Dagan, October 1978).

In Egypt there is relatively slow growth of cultivated area (see table 3.7). Average input in 1960 was 1·7 metres—that is, about 1 metre per crop. Cultivation of cotton (which began with the American Civil War) and of sugar is only possible with year-round irrigation and is therefore dependent on stored water.

Table 3.7. Irrigation development in Egypt, 1882–1960.

	1882	*1907*	*1927*	*1947*	*1960*
Cultivation (million ha)	2·00	2·07	2·33	2·41	2·46
Crops/unit of cultivated land	1·06	1·42	1·56	1·59	1·76

Of the land irrigated in India[27] about half is irrigated from wells and tanks. Of the land irrigated by river flows, the gross water input averages 1 metre in depth: but we must allow for considerable wastage (generally more than half) between the stream and the point of application. A similar figure prevails in the western United States (quoted above).

The 9·3 million hectares of cropped land in Pakistan[28] received

$9{\cdot}2 \times 10^{10}$ m^3 (0·99 metres average depth) canal flow, subject, however, to a loss of 50 per cent in conveyance and application. This average input is too low, and must be contributing to salination. Developments now in hand should raise the gross input to $11{\cdot}4 \times 10^{10}$ m^3. In Israel, where there is supplementary rainfall, the 200,000 hectares of irrigated land receive an average of two-thirds of a metre.[29]

Translation of cubic metres of irrigation water into hectares of irrigated land depends, of course, on the length of the growing season, the rate of evapotranspiration, the extent of supplementary rainfall, and on whether multiple cropping is practised, and on efficiency of application, and so on.

EXTENT OF IRRIGATED LAND

An estimate of the world's potentially irrigable land (other than in North America, Europe, Soviet Russia, or China) was made by the US President's Science Advisory Committee and is shown in table 3.8 below.

Table 3.8. Estimates of global irrigation potential, 1974.

	Run-off from major river basins (10^{10} m^3/yr)	*Arable land (million ha)*	*Arable land in irrigable climates (million ha)*	*Potential irrigation (million ha)*
India	151·2	163	155	75
Pakistan (including Bangladesh)	138·7	35	30	17
SW Asia	6·2	NS	NS	32
Continental SE Asia	85·0	NS	61	10
Brazil	330·0	389	4	4
Tropical S America	67·6	126	6	4
Middle S America	46·4	50	10	4
Southern S America	70·6	108	99	50
Tropical Africa	187·0	389	101	60
Total	1082·7			256

Source: US President's Science Advisory Committee, *The World Food Problem*, vol. 2, p. 447, 1974.

In these areas there were, in 1970, only 82 million hectares under irrigation. It can be concluded that there is still perhaps surprisingly large scope for further exploitation of existing surface waters. However, in the longer term, more reliance will have to be placed upon increasing the efficiency of the management of systems of water conveyance and application which, it will be shown, is generally at present very low.

Data presented in table 3.9 show that in the last 30 to 40 years the area irrigated has more than doubled. In the early 1970s there were more than 200 million hectares of irrigated land.

Table 3.9. Area irrigated in 1930s and 1970 (millions of hectares).

Continent	1930s[1]	1970[2]
Asia	57	163·6
Europe	6	27·7
North America	11	23·1
Africa	4	7·7
South America	3	1·6
Oceania	0·5	1·6
World	81	230·6

Sources: [1] Earson and Harper, *The World's Hunger,* (Cornell University, NY, 1945).
[2] FAO *Production Yearbook,* 1977.

Table 3.11, on page 83, indicates the forty-one countries with more than 400,000 hectares irrigated. China alone has 37 per cent of the total global irrigated area, and the top five countries have 70 per cent. The Chinese domination may be even more striking than this table indicates, because the most recent statistics are for 1960 and considerable development has occurred since that date. However, it is more likely that the Chinese information is suspect and that this estimate is much too high. An estimate by D. H. Perkins[30] for 1957 put the irrigated area at 34·7 million hectares, less than half the FAO estimate for 1960;[31] as shown in table 3.10, Perkins suggests that this represents a considerable recovery and expansion from the 1950 period just after the civil war.

A more recent estimate from an FAO report puts the 1971 irrigated area at 99 million hectares. This is such a large increase that we must suspect that it includes double-cropping. According to the FAO (table 3.11), the whole cultivated area of China is only 111 million hectares, of which 72 per cent is irrigated. At the other extreme, *The Times atlas of China*[32] estimates total irrigated area at 30·8 million hectares. This is quite close to another estimate by Danson of 36 million hectares in 1964.[33] The authors conclude that the available evidence suggests that the irrigated area in China is unlikely to be more than 50 million hectares. (Since 1972 all FAO *Production Yearbook* statistics on China

Table 3.10. Area irrigated in China, 1904–57.

Period	*Irrigated area (million ha)*
1904–09	23·4
1914–19	23·5
1924–29	26·5
1950	16·7
1957	34·7

have included the province of Taiwan which had 500,000 ha irrigated in 1971).

A recent estimate by K. S. S. Murthy, the Chief Engineer to Central Water Commission, of irrigation in India is 44 million hectares, approximately half with medium- or large-scale works, and half with small-scale works.[34]

Table 3.11 shows a global expansion since the early 1960s of about three million hectares per year, two-thirds of it in the developing market economies. In Europe there has been nearly 50 per cent increase with Romania making a spectacular expansion from 207,000 ha in 1961–5 to join those countries with major irrigation systems with 1·7 million ha by 1976. Morocco and Bangladesh have more than doubled their area over the same period. Substantial rates of increases have also been achieved in Cuba, Greece, India, Iran, Korean Republic, Mexico, Philippines, Sudan, Thailand, Turkey, and USSR and, perhaps surprisingly, in view of the latitude, Canada.

The possible confusion over irrigated land and irrigated crops is increasingly a problem in interpreting irrigation statistics. One current trend in irrigation agronomy is to produce innovations which facilitate multiple cropping (for example, short duration, day length insensitive, or cold-tolerant crop varieties). This means that double and even triple cropping is possible. Hence the area of land cropped per year is less valuable than, say, a measure of the net return per cubic metre of irrigation water applied. Any discussion of contemporary irrigation economics has to contain an assessment of the intensity of land and water use and the irrigation system efficiency.

The global status of irrigation intensity is shown in table 3.11. It also indicates the unsatisfactory state of definitions of irrigation area and the poor quality statistics. The original source is again FAO but the global irrigation totals differ considerably from those given in table 3.9.

Table 3.11. Cropping intensity of irrigated land and cultivation of potentially arable land in developing market economies.

Region	Irrigated land cropping intensity (%)	Potentially arable land under cultivation (%)	Irrigated land ($\times 10^6$ ha)
Africa	98	44	2
Latin America	82	28	10
Near East	65	89	24
Far East	132	84	57

Sources: FAO *Indicative World Plan for Agriculture* and the FAO report to World Food Conference, 1974, Quoted by UN World Water Conference Secretariat 'Resources, and Needs: Assessment of the World Water Situation', *Water Supply and Management*, vol. 1, 1977.

The World Bank estimates that the area under multiple cropping is only 10–15 per cent in India and Pakistan (which seems low to the authors), 40 per cent in Bangladesh and 90 per cent in Republic of China.[35]

CROPPING INTENSITY AND IRRIGATION EFFICIENCY

The subject of this section is technical efficiency in the use of land and water. Of course, a much broader economic perspective is also relevant. For example, Ruttan makes the sardonic comment that the productive capacity of the 2·9 million hectares of irrigated land brought into production from 1949 to 1959 was comparable with that of the 10–12 million hectares taken out of production in the same period by acreage allotments and soil-bank programmes.[36]

There can be little doubt that the area irrigated could be extended in most countries if more water were available. In most irrigated areas water is more scarce than land. The Indus Valley is perhaps an extreme example; but the cultivable area commanded by the irrigation works is much greater than the area cultivated. If it is assumed that two crops could be grown on each hectare per year, then the cropping intensity would be 200 per cent per year. The present achievement is about 70 per cent.

A shortage of water at critical times constrains the area cultivated. Tubewell owners and influential landowners near the head of canals have no water shortage. Experience of those owning land in such favoured positions indicates that a cropping intensity of 160 per cent is, in practice, feasible with existing cropping patterns and standards of management. Where land has

Table 3.12. Irrigated land in various countries.

	Land area ('000 ha)	Arable land ('000 ha)	Area irrigated ('000 ha) 1961–65	1971	1976
China	959,696	128,570	77,200	82,100	85,200
India	328,759	164,800	25,523	31,100	34,400
United States	936,312	186,500	14,659	16,050	16,550
USSR	2,240,220	227,400	9,618	11,483	15,300
Pakistan	80,394	19,250	11,139	12,986	13,600
Iran	164,800	15,330	4,800	5,251	5,840
Indonesia	190,435	14,168	4,100	4,490	4,840
Mexico	202,206	26,000	3,700	4,000	4,816
Spain	50,478	15,657	2,089	2,625	2,854
Egypt	100,145	2,690	2,548	2,854	2,826
Italy	30,123	9,364	2,420	2,600	2,820
Japan	37,231	4,415	3,176	2,626	2,690
Afghanistan	64,750	7,980	2,208	2,360	2,460
Thailand	51,400	15,750	1,729	2,106	2,448
Turkey	78,058	24,858	1,336	1,850	2,000
Argentine	276,689	25,000	1,587	1,720	1,820
Romania	23,750	9,760	207	957	1,729
Sudan	250,581	7,450	952	1,300	1,500
Australia	768,685	45,000	1,115	1,470	1,475
Philippines	30,000	5,200	896	1,200	1,430
Bangladesh	14,400	9,180	501	1,047	1,420
Chile	75,695	5,630	1,084	1,200	1,280
Iraq	43,492	5,100	1,150	1,150	1,150
Peru	128,522	3,000	1,041	1,110	1,150
Bulgaria	11,091	3,940	848	1,021	1,147
Republic of South Africa	122,104	13,420	850	1,017	1,017
Burma	67,655	9,514	681	890	984
Republic of Viet Nam	32,956	5,350	992	980	980
Brazil	85,197	29,500	546	830	980
Republic of Korea	9,848	2,060	682	868	936
Greece	13,194	2,923	530	793	867
Cuba	11,452	2,450	456	620	730
Portugal	9,208	3,010	620	623	628
France	54,703	17,139	510	540	565
Syria	18,518	5,260	579	476	547
Sri Lanka	6,561	895	361	439	530
Ecuador	28,356	3,996	446	470	510
DPR Korea	12,054	2,080	500	500	500
Canada	997,614	43,629	365	430	480
Morocco	44,655	7,400	199	355	470
Madagascar	58,704	2,580	306	350	430
All developed countries	5,618,143	648,979	38,666	44,069	50,242
All developing countries	7,777,098	748,755	150,419	168,553	180,314
World	13,395,241	1,397,734	189,085	212,622	230,556

Source: FAO *Production Yearbook*, 1977.

been irrigable but uncultivated, the successful development of expensive up-stream storage such as Mangla and Tarbela dams and the rapid adoption of tubewell irrigation is understandable.

Given unused irrigable land, although it might be expected that efficiency of use of existing irrigation supplies would be high this, in fact, is not so. It is estimated that in the Lower Indus about half the water diverted at the barrage is lost in transit to the plant.[37] Some part of this total loss is evaporated and is irretrievably lost; the remainder joins the groundwater and can be recovered by means of wells, if the groundwater is fresh, but is totally lost if it is saline. In fresh groundwater areas, if the water is recovered, then some water will again be lost by evaporation and some will again recharge the aquifer. In the Lower Indus, over-all efficiency in water use was estimated, after extensive experiments and tests, to be 53 per cent and 80 per cent in saline and fresh groundwater areas respectively.

Irrigation techniques and efficiency

On the Lower Rio Grande in Mexico, Day[38] reports conveyance efficiency to be between 42 and 50 per cent. Across the border in Texas on the same river the lowest conveyance efficiency is 50 per cent; in 3 out of 11 districts studied, the efficiency was 80–2 per cent, and the weighted efficiency was 66 per cent. This higher efficiency results from the recently constructed or rehabilitated distribution network with lined canals or pipelines.

Garbrecht suggests that there are much lower levels of efficiency even in the United States. He estimates only 30 per cent of diverted water is effectively used by crops.[39] In the developing world he considers 40 per cent of diverted water reaches the field but, with only a 30–40 per cent level of irrigation efficiency, the total effective use of diverted water is only 10–20 per cent.

In principle, water should be supplied to meet evapotranspiration needs (for cooling and growth) and deep percolation (for leaching). In practice, this ideal can be nearly achieved with an application system by means of which water is drawn from groundwater, conveyed by pipe to the field, and applied individually to each plant by a slow, intermittent, or continuous drip from a nozzle at pressures which may be adjusted automatically according to plant need. Such a precision system is now standard practice in many glasshouse enterprises and is operated on a field scale for bush and tree crops in Europe, the United States, and Israel. (Costs are $\$_{74}$600 to $\$_{74}$6000/ha, depending upon the number of plants—see chapter 5—$\$_{74}$ is the symbol used in this text to indicate dollars with 1974 purchasing power). In these systems efficiency of water use will approach 100 per cent. Complete efficiency cannot be achieved as there are sometimes small leaks and some water is lost to deep percolation.[40]

Hydroponic techniques have existed from some time. The most recent developments, pioneered by Dr A. Cooper and colleagues at the UK Glasshouse Crops Research Institute, are extremely promising.[41] This is the nutrient film technique (NFT) which consists of growing plants in a plastic or metal trough containing a nutrient solution that is constantly flowing over the plant roots. This nutrient solution is recycled, the level of nutrients is monitored, and the desired balance maintained. Although there are still technological problems, these are being solved and a number of farmers in the UK and elsewhere are developing and adopting commercial systems. The potential advantages that balance the technological risks include nearly 100 per cent water use and nutrient efficiency, no leaching requirement, no soil reclamation or sterilization costs, no human health risks, relatively cheap plastic growing troughs, and very high yields—for example, with tomatoes, large-scale trials approaching 350 tons/ha).

For Australia[42] some shocking figures have been quoted relating to efficiency of water use. From the diversion structure to the farm off-take, losses might be as low as 5 per cent at Sunraysia (pumped supply), but the major gravity distribution systems lost 30–60 per cent (in one case, at Millewa, 97 per cent). Beyond the farm headgate, losses could be 70–80 per cent with 'wild flooding', 30–40 per cent with controlled border flooding, and 20 per cent with sprinklers, even on a calm day. Transmission losses[43] between Blowering Dam and the farm headgate were estimated at 40 per cent. Seepage into the sub-soil[44] from rice lands was put at 2·5 mm/day—of the same order of magnitude as evapotranspiration. But with some sandbed sub-soils seepage rises to the appalling figures of 9–18 mm/day; rice should never have been grown over these, if the soils had been properly surveyed before the irrigation project had been undertaken.

Near Verona in Italy a year-round stream of 'pure' water emerges from a pebbly sub-soil. Its origin has been traced to badly-designed irrigation works in the Po Valley, at a distance of some 40 km.

In some parts of the world, water can be pumped from the sub-soil at comparatively low cost; the Indo-Gangetic plain is an outstanding example. In the humid states of the United States, where irrigation is rapidly increasing, about half the supplies are from sub-soil water.[45] But we still know deplorably little about how rapidly and from what sources these sub-soil waters are recharged. However, in many states, it is evident that recharge is too slow and the water-table is falling. Such mining increases pumping costs. This fact, together with increased energy prices and competing effective demand for water in alternative use is likely to result in a contraction of groundwater irrigation in this area. However, as is indicated in chapter 4, on a world scale the importance of groundwater resources is increasing.

In the Indo-Gangetic plain, the issue of recharge to groundwater aquifers is almost a life-and-death matter. Finney[46] quotes estimates in Sind, that the rate of recharge was as high as 53 cm/yr—nearly all from seepage from fields, canals, and riverbeds. Ghulam Mohammad's[47] estimate, however, for the whole Indus plain, is only 26·5 cm. This is a potential advantage in fresh groundwater areas where it can be lifted and used for irrigation, but such seepage is a major contributor to harmful waterlogging and salinity in saline areas.

In the Philippines[48] downward percolation from rice fields was estimated at only 0·5 mm/day. Horizontal seepage, however, was estimated at 10·9 litres/metre/hour (reduced to 0·17 by the introduction of metal levees). Infiltration into the aquifer can also stem from rainfall. In the high rainfall area of Dacca in Bangladesh,[49] it is suggested that tubewell extraction at an average of 100 cm/yr of water is compatible with recharge, though the most cautious engineers would estimate 45 cm only.

For India, Quraishi[50] estimates that of the rainfall which does not appear as run-off, 86 per cent is evaporated and 14 per cent seeps into the sub-soil—that is, 20×10^{10} m^3 which might be recovered by pumping and which, he suggests, therefore, might irrigate 22 million hectares to a depth of a little less than 1 m, in addition to land irrigated from stream flow. He makes more optimistic estimates than those for India quoted previously—namely, that 55×10^{10} m^3 might eventually be used for irrigation and power.

Efficiency will vary considerably if the water designated for non-agricultural use is high. In France, the densely populated Lille-Roubaix-Tourcoing area of 62 km^2 has an average rainfall of 0·74 m, of which 0·31 m reaches the water-table.[51] Present drawings of domestic water, plus net industrial water in this area taken directly from streams, are about 10 per cent above the national average, and exceed recharge by 10 per cent (or 3 cm of water annually). The water-table has been falling on the average 26 cm/yr since 1921 (about what was to be expected, as the average water content of saturated soil is some 12 per cent by volume).

This, however, was for a densely populated area. It is worthwhile considering the prospects of an urban area, unable to draw supplies from outside, but laid out at the average density of the English new towns—that is, 220 m^2/person of space for all purposes (11,000 persons/gross square mile). We assume that two-thirds of the entire space will be used for public open space and private gardens, covered with grass or other vegetation; also a rainfall of 75 cm, and (if it is not a hot climate) evapotranspiration from the grass and vegetation of 50 cm/yr. The remaining rainfall will therefore appear in the stream run-off or sink into the sub-soil, from which it can be recovered by pumping. One-third of the area covered by structures and roads should yield the full 75 cm rainfall as run-off and 25 cm from the other two-thirds of the area, or 42 cm in all. With the total area averaging 220 m^2/person, this will give a water yield of 10 m^3/person/yr. If industrial water is carefully stored, recycled, and cleansed, such a city should be able to provide its entire water requirements from its own rainfall.

Conveyance and on-farm water use efficiency in most developing countries is low and in urgent need of improvement. Bos contended that water losses due to the complexity of management systems far exceeds losses due to seepage and evaporation from canals.[52] In a subsequent paper Bos and Storsbergen determine from a worldwide questionnaire of the International Commission for Irrigation and Drainage that at field level the staff required per 100 hectares for efficient supervision of water allocation fall from 3·0 for 50 hectare irrigated areas to 1·0 for 300 hectares, to 0·5 for 1400 hectares, finally stabilizing at 0·35 just above 6000 hectares and remaining at 0·35 person per 100 hectares up to 100,000 hectare projects. They suggest the most efficient project size, from a water and manpower viewpoint, is 4000–6000 hectares.[53] Bos is also joint author of an authoritative technical paper on the problems of establishing empirical measures of irrigation efficiency.[54]

Canal lining

Canals may be lined with rigid and flexible linings such as brick, concrete, asphalt, or plastic material. Alternatively, the sides can be layered with less permanent soil or the canal perimeter made less pervious by injections of colloidal material. Local costs of lining vary widely, and there is therefore little merit in citing numerous examples. In any case, the economics of the practice can be established by first estimating losses (for example, by closing a section of canal, maintaining the level with a known volume of water, and deducting estimated evaporation losses). In the second stage, estimates or assumptions are required of the efficiency of the proposed lining material, its life, its cost including maintenance (at market or social prices as discussed later), and the discount rate used to evaluate engineering alternatives. From this information, the discounted total cost of lining a section of canal with seepage losses of, say, 1000 m^3 per year can be obtained. This can be compared with the discounted (marginal) return to an additional 1000 m^3 of water per year over the life of the lining. When these calculations were made in the mid-1960s for conditions in Pakistan no linings were economic.

Further adjustments can be made to the net benefit or cost, to take account of savings through use of greater water velocity and therefore smaller canal cross-section with lining, or additional costs incurred by agriculture if existing canal supplies are disrupted during construction.

Khairpur Tubewell Project was constructed over 150,000 hectares in the mid-1960s. An evaluation *ex postfacto* was carried out by the engineering consultants in 1973.[55] A total of 540 tubewells were drilled, and the first commenced pumping in 1967, 226 in the fresh groundwater zone designed for irrigation, 227 in the saline groundwater zone for drainage, the remainder for either irrigation or drainage depending upon the salinity of the groundwater. The whole project has been working since 1970.

The basic aim of the project was to lower the water-table in waterlogged areas and control it at acceptable levels. This is a pre-condition for reclamation or productive use of additional surface supplies. It appears that the main aim has been achieved and that the water-table has been stabilized at about 3 m below ground level.

There is a marked trend towards use of pipes rather than open lined or unlined channels once agriculture moves to a cash-crop basis. Pipe conveyance reduces seepage and evaporation losses in transit to a minimum, reduces maintenance and, providing the pipes are buried, it enables cultivation to be carried out unimpeded by channels. This latter point is of course much more important in mechanized systems and where land is scarce. The economics of pipelines or other forms of conveyance can be established for any given situation using the logic described for lining and comparing the present value of discounted gains and costs of various options.

Waterlogging and salinity

The twin problem of waterlogging and salinity were briefly discussed in chapter 1. The effects of persistent watering to excess can be very serious indeed. Naylor made an important study[56] of the Khairpur area in West Pakistan. In the old days of irrigation from summer floods only, the fields received an average of 25 cm/yr. With perennial irrigation, the water supply rose to 41 cm in summer and 28 cm in winter; and by 1960, had reached 46 cm and 41 cm respectively, or 87 cm in all. This is more than can safely be put *regularly* on almost any soil without drainage. The soil was becoming waterlogged—that is, the water-table rose near enough to the surface to drown the plant roots, thus making the soil useless. It is estimated that only 2 per cent of the gross water supply was added to the sub-soil each year (40 per cent used by crops, the rest evaporated from uncultivated ground); but this 2 cm/yr of water is enough to saturate 9 cm of soil; and this was the rate at which the water-table was rising.

Spreading water over a wide area (which is often done for political reasons) may lower average infiltration, but it leads to a further evil. Farmers spread the water thinly over their land area, do not leach soils, and speed up the rate of soil salinization, without eliminating the rising water-table. To obviate both waterlogging and salination, Naylor recommended concentrated watering at the rate of 1·2 m/yr with adequate drainage either by power-operated tubewells capable of removing 31 cm/yr, or in some cases open drains, which can remove 21 cm/yr. At ordinary rates of diffusion in the soil, a single well of 2·7 million cm/yr (or 3 cusecs) capacity, working 70 per cent of the time, should suffice to drain 360 hectares. However, because of a lack of funds for canal remodelling, the subsidiary aims of increased surface supplies for reclamation and increased cropping intensity on the drained land have not been achieved. This is a serious loss of potential benefits.

There are often criticisms of the efficiency of public groundwater projects in the Indo-Pakistan sub-continent. In fresh groundwater areas, public and private initiatives have various merits. In saline groundwater areas, there are no alternatives to public initiative, as drainage is a collective good. The Khairpur Project achieved a 40 per cent operating factor between 1971–2 and 1973–4. This included a period when severe floods disrupted electricity supplies. By 1974, out of 540 wells, only 8 have shown serious faults. The main technical problems have stemmed from low voltage supplies to pump motors. The periods of low voltage are attributed to various factors, including the connection of villages to the well-power distribution system.

Water-table control

The salinization of soil surface, which is so serious a consequence of the upward movement of salt-bearing water from the water-table and its subsequent evaporation on the surface, is likely to occur in arid areas if the water-table comes within 3 m of the surface, according to an official American study.[57]

An important factor affecting the rate of salinization in Pakistan is that most cultivators are share-tenants, who are, moreover, moved around by the landlords to different plots each year. As a result, they have little interest in protecting the soil from either waterlogging or salinization. For example, they take little care in field levelling, which is an important prerequisite for successful leaching.

Irrigation project design in arid areas generally includes a leaching component to ensure that sufficient water passes through the soil to depress salt toxicity. In design, this is added as a percentage of half-monthly water requirements, and canals and tubewells are designed to supply this level. Although it is necessary to leach the soil regularly—say, every one or two years—there is no necessity to apply it in even doses throughout the year. As every farmer appreciates that the value of water depends upon the time at which it is available, it is unlikely that they will add between 6 and 20 per cent extra water[58] at high-value periods to leach the soil. They will logically grow 6 to 20 per cent more crop and leach the soil in the months when water has low, zero, or even negative value—such as in the Lower Indus after rainfall or during August, February, or March.

Leaching may be carried out at any time of the year. However, it is an important pre-condition that there should be a low water-table or drainage. When drainage is installed, it is possible to establish the water-table at any desired depth; but, with tubewell drainage, a lower water-table increases pumping costs incurred to lift the annual drainage surplus. With open drains it is mainly the capital cost that will be increased.

Two features which must be considered when establishing a recommended water-table-depth are the effects of a high water-table on growing crops and the rate of soil salinization by evaporation from depth. Some 3 metres were advocated in Pakistan. In the Lower Indus, detailed investigations were carried out on the effect of different water-table levels on the yields of the major summer and winter crops—cotton and wheat. It was found that cotton required a much deeper soil profile than wheat. The optimal profile for both crops varied with the salt content of the groundwater, being much deeper when the groundwater was saline. A relationship between evaporation from sub-soil water and depth to groundwater was established empirically. This indicated that, if the water-table was lowered to nearly 2 metres, 90 per cent of evaporation was eliminated. To lower the table by a further 1·5 metres would decrease evaporation only by 5 per cent. The increased pumping costs of this were capitalized to a present worth, and it was judged that these would not be offset by increases in crop productivity. Following a reasoned assessment of this evidence, it was estimated that 2 metres was sufficient to allow optimal yields of all crops.

It is suggested that this is an important area for further research by a joint agronomic and economic programme. For instance, optimal depth must be deeper in a saline groundwater zone than in a fresh groundwater zone, because the effective root volume is reduced. Furthermore, even the reduced evaporation from depth that occurs with a 2-metre water-table could rapidly cause serious surface salinization without adequate leaching.

Temporary raising of groundwater levels as a result of heavy rainfall would, under saline conditions, prove more serious. In the delta area of Pakistan, storms producing up to 25 cm of rainfall are not uncommon, and they cause widespread crop damage, for they may raise groundwater levels in place 1·7 to 2·7 m—in effect, up to or even above the surface. In these conditions, the saving of capital and recurrent cost by establishing a relatively high water-table would prove to be a false economy. In areas drained by tubewell, it is possible to lower the water-table temporarily in advance of rainy seasons to prevent this occurrence.

Drainage methods

There are two basic approaches to soil drainage: horizontal drainage through open drains or tile drains; and vertical drainage, which is achieved by use of tubewells. A straight cost comparison is not always easy to interpret. If tubewells are feasible, because there is a good-quality (sandy), unconfined aquifer, tubewells are likely to appear cheaper, more flexible and more effective. Where groundwater is fresh the drained water may be re-used for irrigation which is an additional benefit.

Open and tile drains are more expensive because their construction often involves the need both for additional bridges and culverts and extravagant maintenance costs. Tile drains also have very high capital costs. Manipulation of water-table level is difficult once these drains are installed. Controlled water levels can be increased but not lowered; whereas, with tubewells, the system can be easily adjusted by control of withdrawals and the water-table set at any level. As previously noted, if high rainfall is anticipated in a particular season, the water-table can be drawn down in advance so that no damage occurs to crops from temporarily raised levels.

The greatest single problem with open drains relates to the amount of maintenance required because of slumping of the sides and weed growth. In practice, drain maintenance is seldom effectively carried out. In areas with high water-tables and silt soils, it is very difficult to dig open drains below 2 metres. In these circumstances, drains must be closely spaced to perform efficiently—which further increases the costs. Where land is scarce, the fact that open drains taken up 10–15 per cent of the land area must also be taken into account. In some countries (for example, Egypt) drains are important focuses for infections such as bilharziasis; snails which carry the parasite find wet drains an ideal environment.

In areas with high employment, however, it is worth noting that open drains are a potential asset for creating rural jobs. The timing of excavation can be arranged with regard to the agricultural calendar in order to avoid peak periods of labour demand. If shadow-pricing procedures are adopted as a major criterion for testing investment worth, hand-constructed open drains may be the least costly solution. Practical experience suggests, however, that many water-development agencies are unable to solve the financing problems that such a desirable approach presents. Public revenue at the disposal of government is often the most scarce resource, and the opportunity costs of using this revenue for what amount to labour subsidies are considered to be too high. Large-scale projects, which employ many labourers, may build up a big rural purchasing power. If this is not to be inflationary, labour-intensive projects should also be introduced, with ample supplies of basic commodities such as food and cloth.

Salinization

Salinization does not generally arise on rice land. Where it has appeared in Maharashtra it has been successfully treated.[59] Finney[60] takes a more pessimistic view, and discusses rice which has become saline through temporary disuse or contamination by highly saline groundwater. He considers that reclamation of these severely saline soils requires heavy leaching for 1½ to 5 years (2½ to 4 in heavy soils, 3 to 5 years if under *kharif* or summer irrigation only) with efficient drainage. Cropping helps by mobilizing calcium (which must be present) and adding organic matter. Annual inputs (including water losses) of 1·7 m for rice and 1·2 for berseem (a winter clover) are recommended.

Not only the soil but also groundwater and sometimes river water can become saline. The usual measure is total dissolved solids in parts per million (ppm). In practice, the measurement is made of the electrical conductivity of the soil water in millimhos per centimetre—640 ppm corresponding to 1 million per centimetre.

In Australia,[61] cautious limits of salt tolerance have been set for plants, as is shown in table 3.13. Will[62] considers that irrigation increases the salt tolerance of plants. Figure 3.1 shows the greater tolerance of various crops.[63] Barley, sugar beet, rape, cotton, date palm, asparagus, spinach, and various pasture grasses such as Bermuda grass and Rhodes grass are all relatively resistant.

Table 3.13. Limiting values for salt in irrigation water.

	Limiting values (ppm)	
	Total dissolved solids	*Parts of chlorine*
Stone fruit	600	175
Citrus	600	265
Vines	1050	350
Pasture	450	—

In trickle irrigation systems, which are becoming much more popular, the level of availability of moisture below the nozzle is generally constant. Thus soil does not dry out, with the consequent concentration of dissolved salts. In these circumstances, more saline water can be used. However, this increases the cultivator's risks associated with breakdown of the system or other factors promoting intrusion of leached salts into the root zone from deep in the soil profile, which can rapidly have disastrous effects. For example, rainstorms can dissolve salts washed down below the root zone, and a concentrated saline solution can then move into the root area when upward evaporation occurs.

One final source of water resources for irrigation is desalination of saline or brackish groundwater, estuary, or sea water. This

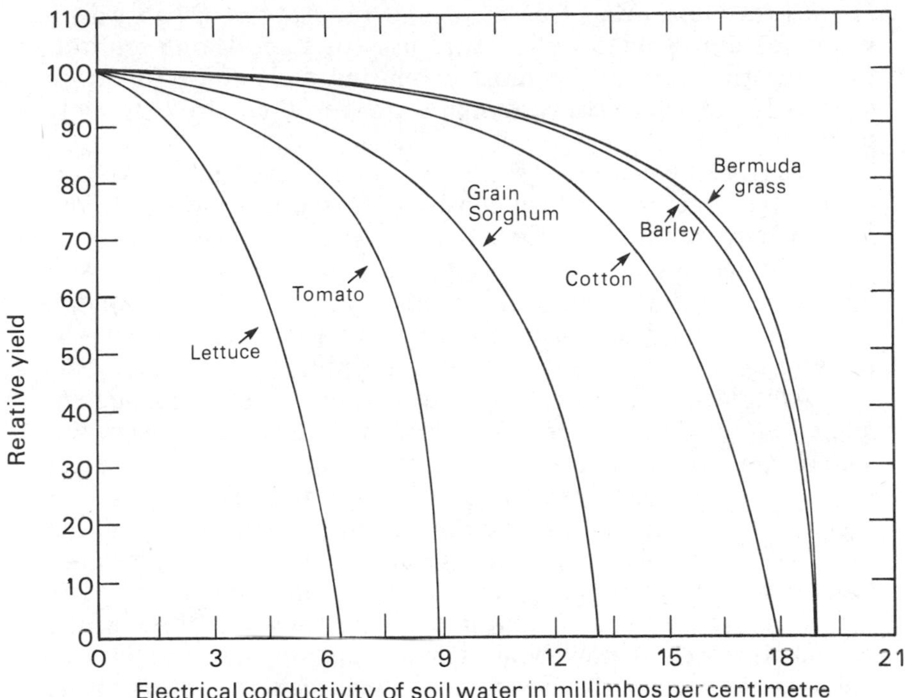

Figure 3.1. Salt tolerance of selected crops.

technology once held great prospects, but the recent energy price rises have dimmed them considerably except for the purposes of obtaining drinking water or for industrial use. However, if cheap energy sources (for instance, nuclear or solar energy) once again appear, then the 97 per cent of global water locked up in the oceans will become the greatest potential water resource.

At present the costs of desalinization (especially sea water with 3·5 per cent salt) are far out of reach of any agriculturalist, except possibly a grower of very high-priced fruits and vegetables in a completely arid country.

In 1977 the most efficient plant could desalinate brackish water at a cost of 26 cents/m^3 and sea water at 80–95 cents/m^3.[64] Kantor gives much higher costs:[65]

	$\$_{76}/m^3$
Multistage flash process	1·00
Multieffect distillation	0·98
Vapour compression	1·05
Reverse osmosis (brackish water)	0·44
Solar distillation	5·00

He also estimates that treatment and recycling of urban waste-water for unrestricted agricultural use costs about one tenth of desalinization and is the most promising technology for areas such as Israel where conventional water resources are extremely limited.

More recently, in the southern hemisphere, the vision of towing huge icebergs, which consist of frozen fresh water, to the areas where water is scarce has come close to reality. In early 1977 one iceberg was cruising in the Antarctic containing sufficient fresh water to satisfy the needs of a city the size of London for 2,500 years. The cause of the renewed interest in what was previously merely a science-fiction concept was the discovery (by two Americans, Dr Weeks (Cold Regions Research and Engineering Laboratory) and Dr Campbell (Geological Survey), who began research expecting to discredit the idea, that the secret was to tow the icebergs at extremely low speeds (0·5 m/sec or about 1 knot). This lowers the power requirement; doubling the speed of a ship raises the power requirement and fuel consumption eightfold. Icebergs in the Southern hemisphere are less jagged, more stable, and float in cooler oceans than those in the north. Research is now being carried out on the feasibility of building large enough tugs to tow huge icebergs, on devising 'plastic bags' to enclose them and to collect melted water, and in contriving handling facilities at the destination.

There are clear prospects of irrigation in the next decade or so in the Atacama Desert of South America, in South Australia, and South-west Africa. The Arabian coast presents problems because the shallow offshore seas will necessitate long pumping distances; even so, the present prospects are more promising than for irrigation for desalinized sea water.

CHAPTER 4 GROUNDWATER ECONOMICS

Groundwater has been exploited by human and animal power for perhaps 5000 years. The ancient Egyptian waterlifting device, the *shadouf*, is still in use, although the application of modern technology, using diesel, petrol, and electric power to increase withdrawals, has been feasible for nearly a century. The wide adoption of modern techniques is a very recent phenomenon. In India, the number of tubewells or boreholes were estimated to be only 19,000 in 1961, but this total had risen to 480,000 a decade later. Similar increases have occurred in Pakistan and Bangladesh. In the Punjab, by 1975 some 9000 wells had been installed in the public sector and about 120,000 by private farmers. In China between 1968 and 1978 1·7 million tubewells have reportedly been sunk in the North China Plain. (Dr J. Nickum, Center for Chinese Studies, University of California, private communication).

TUBEWELLS AND ECONOMICS

The economic importance of groundwater development by tubewells has increased very greatly in the last decade. Tubewells are increasingly used for both public and private exploitation of groundwater. Areas with deficient precipitation and distant river water but with good quality groundwater aquifers will clearly benefit from tubewell development.

Tubewells may be used where surface canal water is at present supplied to augment basic surface supplies and increase cropping intensity. Augmentation to meet relatively short periods of peak demand may be particularly valuable. The most profitable cropping pattern is unlikely to have an even profile of water demand over the year, and a canal system is limited in its capacity to cope with peaks.

In many areas, rivers supplying surface water have a marked

Notes and references for this chapter begin on p. 258.

difference between maximum and minimum annual flows. In this situation surface storage may be built to hold some of the excess water from peak supply periods for release at times of flows. Surface storage is expensive and, if silt loads are heavy, it has a limited useful life. Another possible means of achieving the same objective is to develop the groundwater reservoir which is likely to be associated with the alluvial channel. The alluvial reservoir under the river-bed and active flood plain is pumped during the dry season, and recharged when the river rises in the flood season.[1]

Tubewells have other potential advantages over surface sources. Wells can be sited adjacent to the areas to be irrigated, and therefore costly distribution systems are avoided. If distribution channels are short, it may be economic to line them with impervious material in order to reduce seepage losses. Irrigation systems based on wells can be brought into operation much more rapidly than systems based on a dam and large canals, construction of which may take up to a decade to complete. The importance of early benefits is greater when discounting procedures are incorporated into selection procedures and high discount rates are used.

Tubewell construction can be phased with demand, unlike dams and canal systems. In this way, costly excess capacity in the early years of development can largely be avoided. Indeed, because hydro-dams typically have excess capacity in the early years of operation, groundwater development may sometimes be integrated with low-cost electric poser.

The main disadvantage of tubewells is that they generally have high recurrent costs. If the wells are powered by electricity, these costs can be minimized by using as much surplus or off-peak power as possible. This will often result in night-time irrigation, in which case application efficiency could be maximized because evaporation losses are lower at night. However, in some communities there are strong reactions to night-time irrigation, which lead to poor field supervision. In addition, the lack of ability to see may lead to wastage rates greater than the savings from lower evaporation. Unfortunately, it is generally found that the peak irrigation season coincides with peak power demands, particularly if the the urban sector is well endowed with air-conditioned plant and other cooling devices.

Integration of groundwater with surface water

The extraction of groundwater is complementary to the supply of surface water. It enables the area cultivated to be extended as

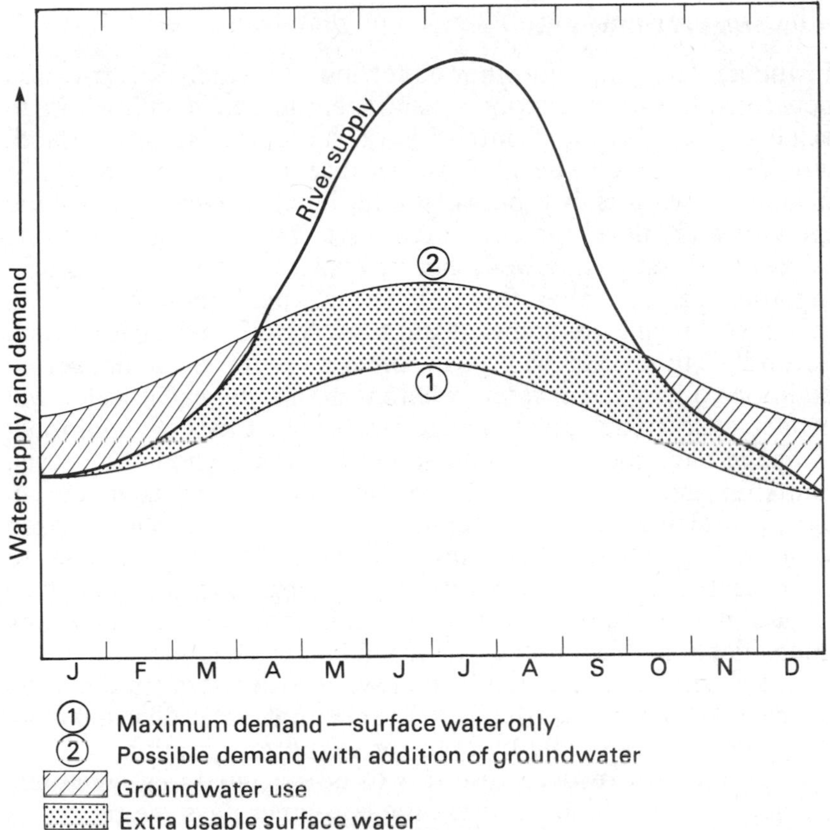

Figure 4.1. Integration of surface and groundwater supplies.

Source: R. F. Stoner, 'Conjunctive use of surface and groundwater supplies' paper to ODI Workshop on *Choices in Irrigation Management*, University of Kent, September 1976.

shown in fig. 4.1 and the peak requirements of high-intensity, profitable cropping to be met with a minimum of excess capacity. It assists with drainage and, if properly operated, allows the maximum input of low-cost surface water without creating a harmful rise in groundwater levels. If canals are remodelled to bring in large quantities of surface water when river flows are high, this will lead to useful recharge. Scarce, low-season river flows could then be diverted elsewhere—say, to areas with saline or no groundwater—and requirements at that time in the remodelled command met wholly from groundwater. This would be worthwhile if the present net value of the value added by the diverted surface water was greater than the discounted costs of canal remodelling, plus groundwater development and pumping.

Advantages of tubewells

In India, Pakistan, and other countries with extensive irrigation networks, the water supply is insufficient to enable cultivators to achieve the maximum intensity of land use. Irrigable land, already developed, lies fallow because of water shortage. If planners have a choice between additional surface supplies and groundwater development, there are two factors favouring tubewells. Firstly, tubewells can be constructed without disturbing the existing canal flows, whereas canal remodelling will inevitably result in some disruption to the existing system. Secondly, tubewells will give some measure of drainage, in addition to irrigation water, whereas additional canal water will in fact exacerbate any drainage problems. Costs of additional drainage have therefore to be added to canal remodelling costs. Shifts in the costs of new schemes and recent advances in agricultural production possibilities (for example, short-season varieties of cereals) have made it feasible and desirable to increase the cropping intensity on existing schemes. Tubewell irrigation and drainage will be an important component of any intensification programme. If groundwater is pumped and re-used for irrigation, efficient use of water will rise from about 50 per cent to 80 per cent (that is, 80 per cent of water diverted from the river will eventually be used for evapotranspiration).

In all areas of resource use it is to be expected that economic efficiency will be highest where the consumer pays the real costs of supplying the various inputs or services demanded. On public irrigation schemes, it is rare for these circumstances to hold. Most public schemes have low, heavily subsidized irrigation rates (see chapter 7). In the case of privately controlled irrigation wells the user faces the full costs, and it is to be expected that he will use water judiciously; that is, he will irrigate until, in his judgement, marginal returns to irrigation will equate with the marginal cost of pumping the well.[2] The fuel consumption over a period of time is a good measure of the water delivered, and this provides the most reliable basis for a volumetric pricing policy; however, this potential advantage will be lost if the surface water is mixed with canal water.

Serious biological contamination of groundwater is rare. Therefore groundwater is an important source of public drinking water in many areas. Generally, no further treatment beyond the safeguard of chlorination is necessary. Drinking water needs are limited. Rural communities in poor countries seldom consume more than 15 litres per head per day if they carry water to the house, or 80 litres per head per day if water is piped to the house. Consequently, wells in a good alluvial aquifer would usually produce a surplus for irrigated agriculture. For example, a typical Indus valley village of 1000 persons would require for domestic use only between 0·4 and 4·0 per cent of the daily discharge of a tubewell. However, where the population have been dependent upon hand pumps for domestic water supply, irrigation projects that lower groundwater table below 6 metres must have provision for year-round drinking water supply and reserve storage.

System components

There are four main components to the tubewell system: the power supply, the pump, the type of screen, and the disposal network. The system can be operated by public or private authorities, or some mix of public and private control. Another major design component is the scale of investment or capacity of the wells.

Power sources

It is not wise to be dogmatic about the optimum combination of each component of the system. Local circumstances dictate the appropriate technology, and the optimum mix will change with time. For example, the main power choice is between electricity and diesel power. Once electricity mains are installed close to the well-field, electric power has a number of advantages. Recurrent costs are much lower, particularly if off-peak power can be obtained. Supervision at the well-head need not be continuous, which will create obvious savings in operator's time.

If an economy is importing oil but has generating capacity for hydro-power, then electric-power has a clear advantage for pumping. Even if oil is imported for all or incremental power demands, conversion of oil into electrical energy at large power stations is potentially a more efficient means of providing pumping energy than direct-drive diesel units. Whether this potential can be realized will depend largely upon the achievement of projected loadings of the power system. The operating performance of generating units and the optimal number and distribution of connections are key variables in achieving optimum loading.

In a publicly developed and operated well-field, electrification can be justified because the timing of installation, location of wells, and duration of pumping is amenable to regulation. In private development, there are, typically, well units that are small and scattered, and there is initial low loading, which creates high overhead costs, thus precluding economic electrification programmes.

Diesel power has several potential advantages to offset the high costs. These include immediate availability, relatively simple technology, and reliability. A diesel engine can often be taken from stock and commissioned in one day, whereas electricity connections are typically liable to long delays. Diesel engines are a well-established technology, even in rural areas, and they are likely to be well maintained. Local manufacture is also increasingly the norm, which has advantages in terms of the availability of spares and the lower foreign-exchange component. Diesel engines (especially the slow-speed type) are a reliable power source. Fuel can be purchased before it is required, thus removing the grave uncertainty with which owners of electrically powered wells in India and Pakistan are familiar. Electric power supplies are erratic, particularly in the peak season. They are liable to fluctuating voltage, which causes problems such as power cuts; they are also increasingly open to labour strikes.

In some areas, the degree of mobility of the diesel unit is an important issue. A small diesel unit can be mounted on, say, a bullock-drawn cart, and taken to several wells or attached to other appliances such as a grain mill. In areas with a high water-table, a cart-mounted pump can withdraw water from several low-cost, small-bore, shallow wells.

In some areas of countries with high population densities and poor farmers, the power sources are animals or even humans. Such technology may appear inefficient, but in fact it is likely to be appropriate to the resource endowment. Reform of such primitive methods should be undertaken only after thorough study of the social as well as financial costs and benefits.

It can be concluded that no general principles can be proposed to aid choice of power source. Detailed investigation and appraisal of local circumstances and opportunities from a public and private viewpoint will be required.

Choice of pump

The type of pump cannot be selected without reference to the power source. Unless electricity is available, deep-well turbine pumps with submersible electric motors cannot be considered. However, vertical shaft turbine pumps can be driven directly from electric motors or diesel engines. Choice can be made largely on the basis of local costings (ideally using economic or shadow prices, as discussed later on pages 214–5). There are some operating characteristics that will influence judgements based solely on costs. Submersible pumps, though cheaper, are more sophisticated, and are therefore likely to be imported and difficult to maintain. Belt-driven centrifugal pumps can be used when the water-table is close to the surface.

Generally, direct-drive diesel pumps will be the best choice for small wells, at the early stages of irrigation development. As the scale and/or experience with irrigation increases, so electric-powered pumps are preferred; first shaft-driven, then the more sophisticated submersible pumps.

In Jordan in 1974 it was found that for highland schemes direct-drive diesel was most economic for projects with up to 10 wells, diesel electric power for 10 to 25 wells, and a central slow-speed diesel power station with high-voltage transmission for well-fields exceeding 25 wells. In the Desert Plateau the conclusion was similar, but because of high salaries the solution was sensitive to the number of operators required to control the diesel units. In schemes with electric power, submersible pumps were the preferred alternative.[3]

Type of screen

A typical tubewell consists of a length of blank pipe pump casing sunk into the ground below the maximum depth to the water-table. The maximum depth is dependent upon two factors. Firstly, during pumping the aquifer may not be able to transmit the percolating water to the well at the rate at which it is being withdrawn by pumping, so the water level in the well falls. Secondly, if the rate of withdrawal for the whole aquifer exceeds the recharge from infiltration, then a temporary or even permanent need for mining exists, which will continue until the increased costs associated with pumping from a greater depth causes pumping to be curtailed and a new groundwater level is established. Determination of the maximum pumping depth is an important function of the design engineer.

The remainder of the well consists of a perforated screen with blank casing used opposite clay bands in the formation. If the well passes through rock, no casing may be used. In the case of larger wells, a gravel filter is placed between the screen or pump casing and the sides of the hole. For a typical large tubewell in the alluvial aquifers of the Indian sub-continent, the hole drilled would be 35 cm in diameter and the screen 20 cm.

A very wide range of materials are used for screen construction. Advanced designs use resin-bonded fibre-glass, which is a material that combines strength and lightness and is chemically inert. This must be one of the few examples of technological spin-off from the space race in use in agriculture. Water which contains salt (say, more than 500 ppm) can lead to corrosion of metal screens (mild steel, for example) and encrustation problems. To combat this, resistant materials, such as stainless steel, brass, or other copper alloys, are often used. However, these materials are more expensive than fibre-glass. In fresh groundwater areas, low-cost mild steel can be used.

All the materials discussed in this section have been used by public authorities for developing groundwater. They generally give a good performance but are expensive in comparison with the alternative local materials which are widely used in the small wells developed by private farmers. Small farmers have exercised considerable ingenuity and have developed low-cost well screens such as coconut coir string wound round a metal frame, tarred hessian on a similar frame, and, most recently, strips of bamboo bound by galvanized iron. In certain rare circumstances, a screen can be dispensed with if a water-filled cavity is developed below the blank casing.

Low-cost screens have a low ratio of open space to solid area, very poor mechanical strength, and a very short life (perhaps 10–15 per cent of that of standard materials). However, there are obvious attractions to a farmer facing a capital constraint. In the Kosi area of India, Clay found that, despite the technical inferiority of the locally developed bamboo screens, they were profitable when combined with shallow depth and a locally adapted, cheap, labour-intensive drilling technique.[4] Use of unskilled labour and lack of foreign exchange made these screens socially as well as economically profitable. Farm budgeting analysis indicated that the low-cost package was more profitable than any technically superior option that was available. This helped to explain the nature of recent irrigation developments in the Kosi area, where there were only 300 iron wells in 1965–6 and no bamboo wells. Bamboo technology was locally developed in the late 1960s, and by 1972–3 there were 19,496 bamboo wells and 3544 iron wells. There was thus a remarkable increase in both types of wells, but clearly the most impressive development was the bamboo option. More recent experiments in 1974 in Bangladesh were a failure because of faulty construction and a fluctuating water-table.

The disposal network

Once water has been lifted above ground level it can be transported to the field by open earth channels, channels lined with various impervious materials, or by pipes. The various linings and pipes require capital investment but increase the physical quantity delivered to the field and lower the pumping cost per unit of water delivered. The most economic means can be determined by comparing an estimate of the present value of lining capital and recurrent costs with the present value of water saved. This follows the same logic as the arguments regarding lining of canals which was discussed in the previous chapter.

In India and Pakistan, the value of water with traditional crops and low yields has been too low to justify capital investment in disposal systems. In some instances it was cheaper to drill two wells and to move the pump rather than build lined channels to distant fields. In cases where fragmentation of land holdings is the rule, there are further complications to be overcome before lining of disposal channels is feasible.

Disposal systems are a neglected area of investment, because they are publicly or jointly owned and the collective interest cannot be enforced. More complicated issues may play a part. For example, prices of inputs used in construction or operation are often distorted, leading to uneconomic use of important resources. A country may be able to produce more irrigation water at the field either by pumping wells or by improved disposal systems. Diesel oil for agricultural use, including tubewells, is sometimes subsidized, whereas raw materials for improving disposal systems, such as cement, may be taxed. Such a pricing policy will lead to more energy consumption and hence expenditure of foreign exchange. Furthermore, unskilled labour, which could be used in constructing improved disposal systems, is typically plentiful and overpriced, and, in the absence of labour subsidies (but subsidized oil) second-best means may be adopted.

PUBLIC OR PRIVATE DEVELOPMENT

Groundwater development is achieved by various types of enterprise. One major area of discussion relates to the issue of public or private investment. There are several important, sometimes conflicting, arguments for each form of development, and choice is often complex and dependent upon local considerations. Furthermore, some mix of public and private development is often optimal—for example, public facilities for distribution of electricity and private wells. In this section, the major general issues are set out.

Advantages of public enterprise

Public wells are the only practical option where saline groundwater is pumped for drainage purposes. Advocates of public enterprise claim that, in areas where there is fresh groundwater, public schemes are the most rapid means of obtaining large increases in the availability of irrigation water. However, speedy exploitation is not necessarily achieved by public schemes. Investigations for the Khairpur tubewell project in Pakistan started in 1959, tubewells started pumping in 1967, but the whole project was not completed until 1970. The project was certainly on a large scale with 540 high-capacity wells producing 50,000 m^3 of water per year, but the gestation period must be regarded as very long.

Private development

In contrast to this, there has been rapid investment in private wells in the sub-continent, where more than half a million new tubewells have been installed over the last decade, enabling irrigation to be extended by perhaps 20 per cent. This progress, though rapid, has not been evenly distributed throughout the area, and there have been sudden spurts of activity or local pockets of investment and whole districts with unexplained inaction. Private development is more unpredictable, and it is therefore difficult to mobilize effective back-up for complementary inputs such as agricultural extension and marketing services. Furthermore, although in principle profitable use of a small tubewell can be obtained by a farmer with 2 hectares or more, and with co-operative grouping even smaller farmers can benefit, in practice only large farmers can command the necessary resources. In some areas this has created disparity in income earnings between large and small farmers.

Private wells have, in addition, two more important practical advantages. Firstly, most governments find it difficult to raise sufficient revenue, and there are many competing demands for the limited supplies available. Private tubewells place only a small burden on public finances (for example, for electricity investments or diesel oil) and may mobilize private resources which would otherwise go into consumption. Moreover, operation and maintenance costs for private wells are borne wholly by the consumers, whereas it is rare for these costs to be met fully in public schemes. In private development, not only government financial costs are saved. In many countries there are constraints in administrative and technical capacity for design, land acquisition, construction, and execution. In planning, because there are undoubtedly alternative uses for these resources, a very broad view of the impact on the economy of public as opposed to private development is required, setting out the opportunity costs of various options. The release of public administrators is a real benefit of private development.

The second advantage of private wells relates to benefits in use. The record of private wells show that in most circumstances they are more reliable than public wells.[5] Water supplies from private wells are 'on demand', and are therefore used so as to be phased with crop demands after crops have obtained supplies from rainfall and cheaper surface irrigation. In public schemes, water is typically pumped into canals and/or taken by a farmer at a fixed time—say, for ten hours every week. It is much more difficult, with this latter system, to ensure that water, and the resources concerned in making it available, are not wasted. Control by farmers over supply leads to high efficiency in use.

General considerations

However, all advantages do not lie with private development. Groundwater is a common property resource; therefore, if individual farmers withdraw what is optimal from their private viewpoint, there is no guarantee that an optimal amount will be withdrawn from an over-all or social viewpoint. Some form of public regulation is essential where groundwater is limited, where there is a risk that overpumping may lead to aquifer deterioration by salt water intrusion, where it is desirable that the water-table be stabilized at a particular level, or where the salinity of groundwater is such that mixing with surface water is required before irrigation. In Pakistan, saline water is described as water at a depth of 300 feet (90 m) with more than 3000 parts per million of total dissolved solids (ppm TDS). Water with 1000–3000 ppm TDS is 'mixing' water, and less than 1000 ppm TDS is fresh water. In the Sind region, sodium content is relatively high and surface water more saline, and so mixing water is 1000–2000 ppm TDS. Mixing water is mixed 1:1 with surface water up to 2000 ppm TDS and 1:2·5 up to 3000 ppm TDS. Regulation of the number of wells by licensing is the most obvious means of control, but this is not always effective as experience at the Wadi Dhuleil well field in Jordan shows.[6] Here, water is being mined despite the best attempts of government to prevent new wells being installed. Control over the amount of water pumped is also necessary. Practical problems of effecting control over private wells are such that it is probably essential to have publicly owned and operated well fields wherever water-table control or the intrusion of salinity are actual or potential problems.

Wells in fresh groundwater areas that are publicly developed and operated may enable government to withdraw scarce surface water for use elsewhere in the country where groundwater is absent, but the needs are greater or returns to marginal supplies are more substantial. The incidence of groundwater development costs should be on the beneficiaries—in this instance, the farmers receiving the diverted surface water.

The case for public development can be based on the often declared intent to aid the most needy groups of farmers. These may be small farmers in a relatively rich or high developed area, or whole regions may be backward. Wells require large sums of capital investment and are therefore out of reach of very small farmers. Clay estimates that the minimum holding size required to provide sufficient to earn a 10 per cent rate of return is 1·6 hectares for a brass screen or 1·0 hectare for a bamboo boring plus a pump set. In fact, because of problems such as farm fragmentation, risk premiums, and so on, the minimum holding size is normally 4 hectares.

In these circumstances, public enterprise may be justified to ensure equitable distribution of groundwater to all holdings. Wells may be retained by government, or, as in Bangladesh, given to a co-operative to manage distribution. Private farmers may also carry out a similar function by selling excess water to their neighbours. In parts of India another practice is growing in importance. Farmers may own wells but hire a portable diesel pump set which may be owned by a (landless) contractor. In this way the problem of a large indivisible investment is at least partially overcome. Unfortunately field studies in several countries reveal that neither public nor private development of groundwater can ensure adequate and equitable distribution to all farmers, large and small.

Private tubewell development, though rapid in many areas, has been erratic. There are some poor districts that have, apparently, ecological conditions similar to other highly developed districts, but where, for one or more unknown reasons, groundwater development has not occurred. In these areas some form of public development may be justified, such as a few public wells to serve as a demonstration to potential private developers.

The technology debate is increasingly seen as being linked to other more general problems of society. For example, in Bangladesh modern deep tubewells have been monopolized by large landowners.[7] They have been sited, in many instances, not to satisfy technical or social goals but to suit locally powerful individuals. They are also operated in large by their interests and in the process give rise to opportunities for additional profitable activities. For instance, once reliable water is available, they may be able to outbid competitors in any land purchase or land leasing transactions. An alternative technology of small-scale tubewells or hand-dug wells would not be so readily controllable by a few individuals. Furthermore there is evidence that the powerful farmers pump wells only for their own use, fearing that prolonged pumping will lead to a breakdown which will not be repaired for an inordinate time. This is certainly a reasonable expectation. Even the planners in Bangladesh expect 57 per cent of tubewells to be inoperative at any moment.[8]

In these circumstances, deep tubewells, which are a technically superior means of obtaining groundwater, become inappropriate given the socio-political realities. It is important, however, to recognize that it is the inadequate institutional development which makes the technology inappropriate.

DESIGN PRINCIPLES[9]

In designing wells, the engineering objective is to produce water at the field at least cost. In economic analysis, a distinction is normally made between economic and financial cost concepts (see chapter 5). In this chapter general principles are used that are valid for either approach.

Engineering considerations in costing

Total costs are made up of capital investment costs and recurrent operation and maintenance costs. The system with minimum costs can be determined by conventional discounting techniques.[10] The key variables in design which affect costs are, typically, well screen length, well screen diameter, design discharge level, and drawdown.

Capital costs are a linear function (except for very short lengths of screen when it becomes necessary to lengthen the pump casing) of these variables:

$C = F_1(D, W, d, L)$

where
C, capital cost
D, depth of well
W, depth to water level during pumping (1)
d, diameter of well
L, length of well screen

Recurrent costs are a non-linear (quadratic) function:

$c_n = F_2(W, L, Q, m, T)$

where
c, recurring cost in year n
Q, discharge (2)
m, maintenance costs
T, hours pumped per year

For any well design, the present value of total costs can be determined by estimating

$$\text{present value } PV = C + c_1 f_1 + c_2 f_2 \ldots c_n f_n \tag{3}$$

$$f_n = \frac{1}{\left(1 + \frac{r}{100}\right)^n}$$

n, life of project in years (4)
f_n, discount factor as shown
r, discount rate as percentage

The optimum value for any parameter can be determined for any given value of other variables by partially differentiating PV (equation 3) and equating to zero. Thus $d\,PV/d\,L = 0$ would give optimum screen length for a given discharge, diameter, discount rate, and time of pumping. By setting the first differentials of each variable to zero in turn, the relationships can be explored and the optimum design for each set of conditions determined. In this way the sensitivity of the designs to a range of plausible assumptions can be tested.

Stoner *et al.*[11] used this procedure for Bangladesh conditions in the late 1960s, and found the expression for present value for a given discharge (57 litres/sec) and annual running time (1200 hrs/hr).

$$PV = 28{,}863 + 115\,L + 3231{\cdot}5\,Q + 158{\cdot}75\frac{P\;Q}{L} + 288{\cdot}5\frac{P\;Q^2}{L}$$

PV, present value in rupees
P, a constant relating to aquifer permeability (5)
Q, discharge in cusecs
L, length in feet

From this expression the optimum screen length L can be determined to be 24·7 metres. The authors then conducted a sensitivity analysis of well depth, varying assumptions on annual operating hours and well capacity and sub-optimal screen lengths. These showed that, as would be expected, the lowest cost per unit of water pumped comes from a high-capacity well with a long screen and a high use factor. However, for a given discharge, the total cost curve is very flat in the region of the minimum.

This feature is illustrated in fig. 4.2. The flat total cost curve shows that screen length is a relatively flexible component of total costs. This has two important practical implications. Firstly, the optimum depth will change with changes in discount rate; the higher the discount rate, the shallower the well. A high discount rate has the effect of favouring substitution of recurrent cost expenditures for capital costs. Secondly, the flexibility opens the possibility of shifting the incidence of costs to various components of the investment without changing social costs. Because in a typical low-income economy there is a high opportunity cost for capital, from this viewpoint a high discount rate and shallow wells are optimal. However, low-cost capital aid may be forthcoming which would favour deep wells. In some instances recurrent revenue is more difficult to raise than capital aid, and this would also favour deep wells. Some governments find it operationally useful to adopt the (arbitrary) regulation that recurrent costs must be met by users. In this case (and in the absence of external finance), the lowest incidence of public support will come from a shallow well (low discount rate).

Stoner then conducts a similar exercise for optimum discharge using equation (5), and concludes that:

> if there is enough aquifer available then it is cheaper to go for the optimum screen length for whatever discharge is chosen or fixed. If, however, the screenlength is fixed say by the amount of aquifer available, and the objective is to obtain the cheapest water possible, then the optimum design capacity for the fixed screen length is required.

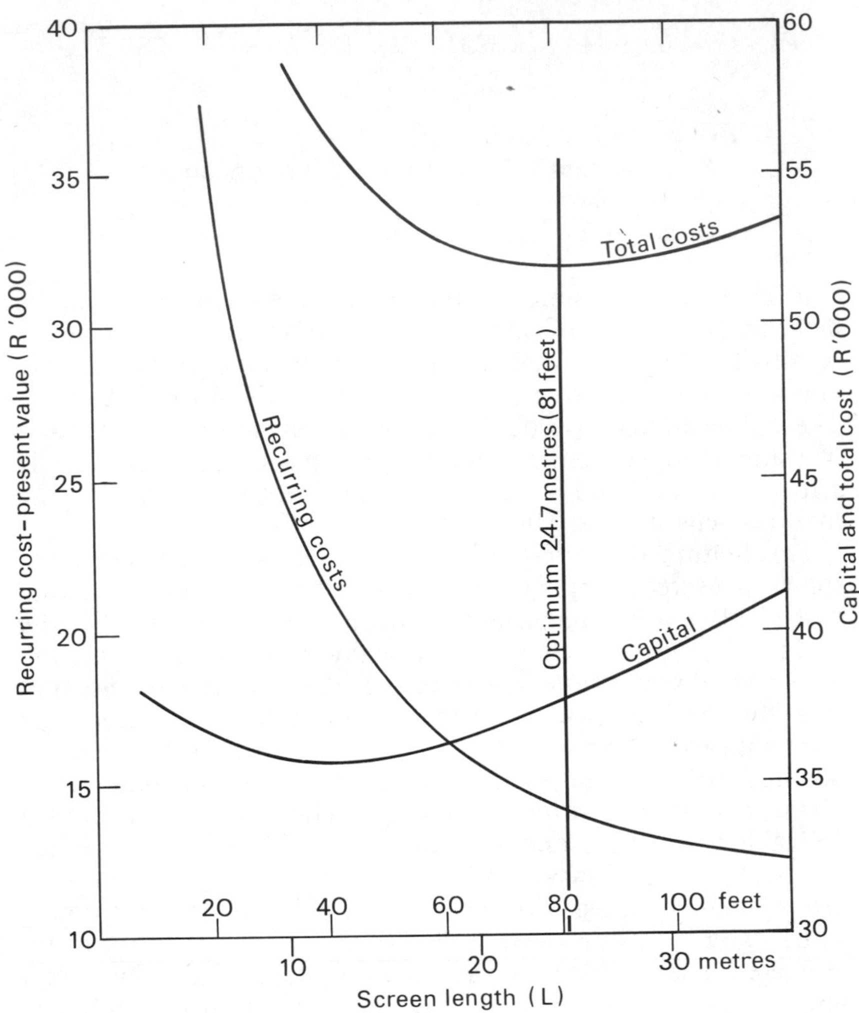

Figure 4.2. Optimum screen length.

Furthermore a conclusion:

> that is constantly recurring in economic analysis of groundwater development is that the cost of water delivered at the well head steadily diminishes as the capacity of the well increases, even allowing for the fact that the wells are all of optimum design and that this means a much higher initial cost for the larger capacity wells.

As with any piece of equipment, the lowest unit water costs for any given well will be obtained if high load factors are maintained.

Other factors in costing

In the section above are given engineering conclusions which relate to the cost of water at the well head. Other considerations may be important. For example, it was previously indicated that distribution losses will rise with an increased area served; farmer control is desirable, and this may require smaller sub-optimal capacity wells with higher benefits in use. Low load factors may be economic if there are high benefits from short-season irrigation (for example, in Bangladesh there is only a four-month dry season). Thus it is not possible to optimize a well system unless the set of benefits which match a particular design are considered simultaneously with costs.

Variation in diameter of the well is also worth consideration, because a large-diameter borehole is not proportionately more expensive than a small borehole. Drilling costs in alluvium are roughly constant, and screen costs are given by the relationship 11 NM for glass reinforced plastic and 5 NM for mild steel pipe

where N is a constant
and $M = 4{\cdot}9 + 1{\cdot}7\,R + 0{\cdot}025\,R^2$
with R = radius of screen or casing in cm

The results of sensitivity analysis on diameter show a similar pattern to depth in that the total cost curve is very flat near the optimum. A large-diameter well also enables capital costs to be substituted for recurrent costs. In Bangladesh, the optimum for glass-reinforced plastic is between 15 and 25 cm.

Deep wells in confined aquifers, such as are found in Saudi Arabia, have their length of screen fixed by aquifer geometry. Since screen is a small proportion of the capital costs, the optimum length of screen will be near the maximum possible. However, diameter could be varied, and Stoner has found that the diameter of 17 cm used in Saudi Arabia was sub-optimal. It was derived by convention from oil-drilling technology, but each litre pumped costs 33 per cent more than that from a screen only 5 cm greater in diameter.

This essentially technical view of the optimum well may be inappropriate when the full range of social, political, and economic factors are considered. J. W. Thomas, in a well-researched and well-argued case, makes the following points about tubewell development in Bangladesh.[12] He defines three categories of well from the large number of options given choice of drilling technique, power and engine source, pump, screen, and institutional support. These categories he terms high-cost, medium-cost, and low-cost wells. In table 4.1 the main results of his comprehensive analysis showing favour to low-cost wells are evident. As this refers to an early stage in development of groundwater, the government's request for aid for equal numbers of high-, medium-, and low-cost wells may be appropriate. However, there is in fact a strong bias in the request towards high-cost wells, given that the cost differences were in the ration 6:2:1 respectively. He shows that besides the economic returns being most favourable for low-cost wells, the consistency with declared government development objectives, the employment creation and training, the distribution pattern of benefits, and the domestic industrial impact were all, in practice, more satisfactory with the low-cost wells.

Table 4.1. Choice of technology for tubewells in Bangladesh

Category	*No. requested by government*	*Internal rate of return*		*Committed*
		market price	*shadow price*	
Low-cost	1500	48	51	None
Medium-cost	1500	33	25	9000
High-cost	1510	7	4	800

Source: Thomas, *op. cit.*

Why, then, it must be asked, was such a considerable advantage ignored? The answer is complex. The facts of over-valued exchange rates, wage rates set institutionally higher than the real cost of labour, and a predilection by agencies and donors of capital aid for the 'modernity' associated with medium- to high-cost wells provides a partial explanation. The donor preference stems largely from technical judgements on speed of installation and operating efficiency, on the fact that the economic return to the familiar medium-cost wells was satisfactory, on the larger number of unknowns in the design and operating performance of low-cost wells, on the difficulty of controlling a much more scattered programme of small drilling rigs and small wells, and on the mistaken judgements regarding the relative difficulty of operation and maintenance of the alternative systems.

Biggs, Edwards, and Griffith go further.[13] They claim that manually operated hand pumps can irrigate 0·2–0·3 ha paddy at a capital cost per hectare 60 per cent higher than shallow tubewells, but with the undoubted advantage of using village skills and equipment for operation and maintenance, no fossil fuel, no dependence upon government for installation or maintenance (indeed there is a 40 per cent tax on capital cost), and no reliance upon co-operation among neighbours in operation. The hand pump is suitable for smallholders because of relatively low cost, and, very importantly, given massive land fragmentation, the main pump unit is mobile. However, this irrigation requires nearly 500 man-days of labour per hectare, and, although this is splendid in providing employment and the gross return might be as high as 5 kg rice per man-day, the variation is likely to be substantial and field-determined energy budgets might produce a less favourable picture.

Hamid *et al.*, Thomas, and Biggs *et al.* give plenty of ammunition to those who believe that development planning is all too often a conspiracy between aid donors, bureaucrats, consultants, and big farmers, to retain or expand their business whilst extracting surplus from the small farmers. However, a more charitable interpretation is that our understanding of complex social systems based upon irrigated agriculture is still inadequate for definitive interpretation. Although natural science and economics have important roles to play in improving our perceptions and decision making, such narrow perspectives will not prove sufficient to the formidable task of improving irrigation farming; in fact, they may, in many instances, be misleading.

MINING OF GROUNDWATER

There are two elements to be considered in designing a tubewell project to exploit fresh groundwater. The first is a stock resource which is the water that has been built up slowly over a long period and is stored in the soil reservoir, and the second is a flow resource which is the annual addition to groundwater by seepage from rivers and canals and deep percolation from fields after rainfall or irrigation. Where recharge is nil or negligible, as in Saudi Arabia, the only option open is irreversible exploitation or 'mining' and decisions are restricted to the rate of withdrawal. Where, however, recharge is significant it has been proposed by Renshaw[14] that the value of increased pumping cost resulting from mining, say 1000 m^3 of water (which increases pumping costs in perpetuity), should be capitalized and compared with the benefits obtained from irrigating with this 1000 m^3. Where recharge is important it is clear that the extra costs involved to pump all this water through the extra height will be considerable. In any project area the running costs of the proposed wells can be estimated using the expression:

\$/well/annum = kQHF
where Q = well discharge cumecs
H = height water pumped in metres
F = well operating factor (proportion of time operating)
k = a constant

If 1000 m^3 in excess of recharge is lifted, the water-table will fall by the reciprocal of the fraction of soil occupied by withdrawable soil water. Thus if one-fifth of the aquifer is withdrawable water and 1000m^3 per hectare in excess of recharge is withdrawn, the water-table will fall five centimetres and all recharge must be pumped the extra distance. One crop of cotton in an arid environment, with soil water storage capacity such as is found in the Indus Plains of Pakistan, would cause the water-table to fall by 50 cm. In this event all future recharge must be pumped through the extra height in perpetuity. In assessing the economic value of this crop of cotton the discounted present value of the increased future pumping costs, plus the cost of lifting the water used to irrigate the crop under consideration, have to be compared to the marginal value of the cotton crop.

In 1967 in Pakistan the solution to the mining problem was found to be extremely sensitive to the discount rate selected. A high discount rate favours mining because the value of the crop grown is offset by increases in future costs which receive less weight. Given the returns to cotton that obtained at that time mining became an economic proposition at a discount rate of 15–20 per cent. This was very close to the expected returns in the public sector. However, any increase in the net returns to cotton (by say a yield increase or an increase in the foreign exchange value of cotton exports) would favour a decision to mine water.

Unfortunately a mining policy is seldom arrived at on the basis of a rational forecast and subsequent calculations. Where uncontrolled private groundwater development occurs it will pay no individual to limit pumping to retain groundwater table at an economically optimum level. Even when a safe yield has been achieved an individual will gain the total return from his own pumping less the cost of average damage that his pumping inflicts upon his neighbours. The situation is analogous to that which leads to overgrazing of common grazing land when numerous farmers have access to it. Freedom of access in a common eventually leads to ruin for all. This is as true for groundwater development as for grazing resources.

The water-table is falling in many countries as a consequence of its unregulated exploitation for irrigated agriculture. It is falling not only where there are highly productive agricultural systems with strong and growing competing demands from industry and urban demands such as in the Western States of America. In Wadi Dhuleil in Jordan tubewell yields are decreasing and in some circumstances wells have gone dry (Clayton *et al.* ibid.). In India where village electrification is receiving government emphasis Kazmi estimates that a 1 per cent increase in village electrification results in an additional half-to two-thirds of a percentage point increase in the rate of utilization of groundwater.[15] It is therefore hardly surprising that in the same journal there is a report from four districts in Tamil Nadu indicating a fall in water-table resulting in a five-fold increase in pumping costs. In Coimbatore the water level is reported as to have fallen 60 metres. In many countries the success of groundwater development will inevitably increase the problem of mining and the need for public control of withdrawals (and possibly distribution) to ensure they remain at socially optimal levels.

CONCLUSION

It is concluded that groundwater development is likely to become a more important activity in the future and that the design issues and institutional arrangements discussed in this chapter deserve further theoretical development and empirical verification. Establishing the costs and benefits of alternative means of integrated groundwater development is a complex but vital area of study in many developing countries.

CHAPTER 5 THE COST OF IRRIGATION

GENERAL REVIEW

The cost of irrigation water may vary greatly. In previous editions of this book great stress was placed upon establishing relative costs of irrigation facilities within and between countries. In this edition, most of those data and additional information are presented. All the costs have been updated to 1974 values of purchasing power ($\$_{74}$), using an index of international prices prepared by the International Bank for Reconstruction and Development (IBRD) and set out in table 5.1.[1]

There are methodological and empirical problems with all indices. Therefore the reader is advised to exercise judgement interpreting the following data. There are three important reasons for this. The first relates to the price index. Any composite price index is suspect, and caution is necessary when comparing costs over long time periods, from several countries, during periods of fluctuating exchange rates, and differential inflation of factor costs. Secondly, it is well known that the prices of inputs used in irrigation development are often distorted for various reasons such as government tariffs, taxes, or subsidies. It is seldom possible, using secondary sources, to distinguish and make adjustments between real economic costs (or shadow prices) of irrigation, and apparent or financial costs. The importance of this distinction is discussed in the final chapter. Thirdly, irrigation projects are increasingly part of multi-purpose developments, and therefore any costing requires arbitrary allocation of joint costs (for example, the cost of a dam between irrigation, hydro-power, and downstream flood protection).

Prices have been converted into dollar values at the exchange rate prevailing in the year in which they occurred, using rates published in the *UN Statistical Yearbook*[2] unless a different rate is specified. The dollar values are brought to 1974 levels using the index in table 5.1. The dollar exchange rates for the main

Notes and references for this chapter begin on p. 259.

Table 5.1. Index of international prices ($US).

1960	54·9	1965	57·8	1970	64·5
1961	55·4	1966	58·8	1971	69·3
1962	56·0	1967	59·2	1972	75·9
1963	56·3	1968	58·9	1973	88·9
1964	57·2	1969	60·9	1974	100·0

Source: Economic analysis and projections department, IBRD., 1974

countries studied are shown in the Appendix. Despite the imprecise nature of the final '1974' values, the data presented in this chapter are a valid indication of the approximate order of magnitude of costs, the wide range within and between countries, and the tendency for long-run marginal costs to rise.

Comparisons among countries of the purchasing power of money show very great variations among commodities; in a poor country, personal services are relatively cheap, and cars, for example, relatively expensive. In irrigation, the costs concerned consist principally of the purchase, maintenance, fuelling, and lubrication of machinery, and construction costs, not of buildings but of dams, channels, drains, soil levelling, and so on. Comparisons among the USA, Western Europe, and Latin America in addition to other information, show that the exchange rate is, after all, a fair indication of comparative costs in these fields.[3] However, for the rest of the world (including Australia and New Zealand) the best indication is the exchange rate multiplied by 1·25—taking account of the fact that in these countries costs are 25 per cent higher than indicated by the exchange rate with the USA due, among other factors, to transport and installation costs on imported equipment and protection of local manufactures.

The international price comparisons given by Kravis cover 10 countries with a great range of real incomes per head. For the prices of equipment ('producers' durables'), although there are individual fluctuations, there is no tendency, in descending the income scale, for comparative prices to depart from the exchange rate ratio. It is quite otherwise, however, with 'construction excluding buildings', which even now is apparently labour-intensive, and whose costs are much lower in low-income countries. On a US base, costs are only 42 per cent of the exchange rate in Italy, 24 per cent in Colombia, and 18 per cent in India. The exception is Kenya, at 83 per cent (attributed to an extreme shortage of carpenters).

On the output side the situation is similar in almost all poor countries. Returns are under-valued, because the scarcity value of foreign exchange is greater than the exchange rate indicates and wages tend to be higher than opportunity cost of labour. Therefore, agricultural returns, in terms of imports saved or exports promoted, are under-valued. Corrections have not been made to the costs and benefits in this and the following chapter to reflect imperfections in exchange rates, but the reader should be aware that, in general, costs for Africa and Latin America tend to be under-estimated. When net returns are considered, the under-estimated costs and benefits in Africa and Asia will tend to cancel each other out.

PRIMITIVE PUMPING

Not only the first example of irrigation but also, to the best of our knowledge, the earliest-known practice of any form of agriculture took place about 7000 BC at Jericho, a favoured spot where water from rock springs emerges in a hot, dry climate.

Even more favoured, however, are those areas where the natural summer rise of great rivers, flowing through flat country, can be used to provide 'basin-irrigation' for large areas, at the cost of no more effort than is required for channelling and bunding (separating the fields with small dikes). Such irrigation provided the food supply for three of the world's most ancient civilizations, those of the Egyptians, of the Sumerians (on the Euphrates-Tigris flood plain), and of the Indus Valley. (Whether these civilizations developed their ideas independently, or were in communication with one another, the archaeologists are still unable to tell.)

The next stage of difficulty arises in countries such as the Sudan or Bengal, where water is available in river-beds but 2 or 3 metres below the level of the cultivable land. From ancient times, simple devices have been used, depending on human labour, for raising water. More advanced devices, using animal power, can raise water from river-beds or wells, up to what appears to be an economic limit of about 9 metres' depth.

The *shadouf*, to use the Arabic word, or swinging bucket with counterpoise operated by one man,[4] will deliver 3–5 m^3/hr from a depth of 2 metres (6–10 ton-metres/man-hour). Other devices[5] for use at depths of less than 3 metres, are the scoop, swing bucket, Archimedean screw, paddle wheel, tympanum, water-ladder, and *mhote* with only two buckets. Dias[6] similarly finds that a large well-sweep worked by 5 men will deliver 6 m^3/hr from a depth of 9 metres (10·8 ton-metres/man-hour). Tainsh, on the other hand, expects only 3 m^3/hr from the same depth for a *piccottah* (large counterpoise) worked by 4 men (6·7 ton-metres/man-hour). Other devices used for depths of 3–9 metres are the Persian wheel (*noria*), the Roman wheel (sometimes water-powered), and the *mhote* with a chain of leather buckets (in exceptional circumstances this has been worked to a depth of 100 metres). Dakshinamurti *et al.*[7] suggest much higher rates of average discharge: for the *piccottah*, 8–11 m^3/hr; for the swing bucket, 14–19 m^3/hr; for the Archimedean screw, 14–15 m^3/hr; for the Persian wheel with pair of bullocks or a camel, 14–18 m^3/hr.

Sansom[8] described water-wheels in Vietnam, where irrigation is from ditches with the water-table near the surface, lifting water for 0·5 m only, powered by four men, and delivering about 7 m^3/hr, a very poor performance. Oxen tire more quickly than men, and a pair of oxen can generally work only 5 hours a day. However, working a *saqiya* in the Sudan,[9] when the Nile is high and the lift only 2 metres, a pair can irrigate 1·8 hectares. Assuming 10 mm/day evaporation, this implies an output of 36 m^3/hr, or 72 ton-metres/hr, over 7 times a man's output. Output fell to 20 m^3/hr in winter, and to 10 m^3/hr in the hottest season, when the Nile was lowest and the lift greatest. One horse-power is officially defined as 272 ton-metres/hr (550 foot-pounds/second). Even after making a substantial allowance for frictional losses, it is clear that a pair of bullocks fall short of 1 horse-power—which most horses cannot in fact attain. Horses were originally bred for war; their use in agriculture would have required too much capital investment. The replacement of oxen by horses as the principal source of draft power in agriculture did not take place in England until the eighteenth century, and in most of Asia—and, indeed, in some parts of Europe—has not yet taken place. Oxen were still being used to haul timber in rough country in Australia as late as the 1940s.

Similar results in the use of man- and ox-power are shown by Dias and Tainsh (who are in agreement) for two men working with two oxen, whose output (from a depth of 9 m) is 7 m^3/hr with the *mhote*, and 9 m^3/hr with a Persian wheel. These represent, respectively, 63 and 81 ton-metres/hr, or 8–10 times a man's output, confirming that a pair of oxen, in simple work output, is equivalent to about 7 men.

Although we know the wages of rural workers in India—and the fact that they are inordinately low, probably substantially below the true marginal product of labour—there are wide variations in the price of bullock labour, in terms of human labour. In his production function for the village of Senapur, Hopper[10] makes the marginal value of 1 hour's work by a pair of

Table 5.2. Costs of simple lifting systems in Uttar Pradesh, India.

	Depth (metres)	*Supply m^3/hr*	*Capital cost $\$_{74}$*[a]	*Average life (years)*	*Interest depreciation & maintenance $\$_{74}/yr$*	*Interest depreciation & maintenance c_{74}/hr*[b]	*Interest depreciation & maintenance c_{74}/m^3*
Swing basket (large counter-poise)	1·2	16·5	1·63	2	0·98	0·065	0·39
Bucket	6	2·25	10·85	5	3·32	0·22	0·98
Leather bag	9	7·2	13·56	5	4·14	0·28	0·39
Persian wheel	9	11·3	189·9	10	38·9	2·59	0·23
Earth well or tank	—	—	68·0	10	14·6	0·98	
Brick well	—	—	403	75	40·0	2·67	

a. 1963 rupee = 27·1 c_{74} b. assuming 1500/hr/yr

bullocks, inclusive of the labour of the driver and the use of the equipment, no less than 9 times that of a man-hour. This, however, was in a region with a short monsoon season, during which the demand for bullock labour for cultivation was extremely urgent. Dhondyal[11] valued a day's work by a pair of bullocks at about 5 times a man's wage in Uttar Pradesh in 1963–4. A careful study in the Punjab by Shastri[12] showed that the full cost of a bullock, inclusive of interest and depreciation, was Rs558/yr in 1950, and that it could usefully be employed only about 160 days/year; even the largest farms could not achieve more than this—or Rs3·5 per working day. At Punjab rates, this was about 7 times a man's wages for a pair of bullocks. In West Bengal,[13] where oxen appeared at that time to be abundant and the rainy season prolonged (so that there is less urgency about cultivation), the average agricultural wage worker was paid at the rate of 3·5 kg rice/day, and at only 6·6 if he provided the plough and a pair of oxen. Bullocks are economic in areas of low or seasonal rainfall with more uneven labour demands and quite sharp peaks at critical times. It appears also that the number of oxen in Bengal was greatly in excess of the number required for cultivation.

The horse is not a ruminant (four-stomached) animal, and although it can eat a certain amount of grass (having an exceptionally large caecum), it requires substantial grain feeding, certainly if it is to do any work. The ox is better equipment for a poor farmer, because for much of the year it can be fed on straw and refuse, though it too requires grain-feeding if it is expected to work hard. (It may be surmised that the low-paid, over-abundant oxen in Bengal receive less food and work more slowly than elsewhere.)

In parts of the Punjab the average size of area cultivated depends on the quantity of water which two bullocks can draw from a well, which in turn depends on the level of the water-table. Capital costs were estimated by Dhondyal (*op. cit.*) for Uttar Pradesh in 1963–4—though his estimates of operating costs appear far too high (see table 5.2). Wells sunk in the Punjab at various dates in the past to depths in the 6–14 metre range, with average width at month of 2·5 m, show a much higher average cost of $\$_{74}$1550. Even if they worked for 1500 hr/yr and deliver 9 m^3/hr on the average (both optimistic assumptions), at 10 per cent interest rate this represents a capital cost of 2·09 c_{74}/m^3.

Operating costs estimated for the 1950s on the basis of 1R/day for labour and 5R/day for a pair of oxen (converted at 34·9 c_{74}/rupee) were as follows:

	c_{74}/m^3
Piccottah	4·7
Well sweep	2·9
Mhote	6·0
Persian wheel	4·6

Capital costs, it will be seen, were small in relation to operating costs. It will also be observed that even with such low-paid labour these costs are very high in comparison with those from power-pumping, in India or elsewhere.

It has, however, been suggested[14] that a simple application of bicycle pedals to a Persian wheel might raise a man's productivity to 2 m^3/hr (from a limited depth)—that is, a cost of only 2·2 c_{74}/m^3. Water drawn from 12-metre depth by oxen in southern Italy in 1913 was costed[15] at 9·3 c_{74}/m^3; clearly, this could be used only on valuable crops.

The cost of water obtained by a manually operated water-wheel in Vietnam,[16] where real wages are far higher than in India, amounted to over 15·5 c_{74}/m^3. This water was used for growing vegetables, for which US Army camps provided a high-priced market. There has, however, been indeed a strong incentive to substitute mechanical pumping.

In Bangladesh, opposite forces are at play. This country has a cultivated land area of only 0·12 ha per person, a population growth rate greater than 2·5 per cent per year, and an acute foreign exchange shortage. In the face of abject poverty, many farmers are being forced to use low-cost hand pumps delivering less than 2 m^3/hr to irrigate rice fields. This is technically inefficient but economically viable, given the low opportunity cost of labour and the high opportunity cost of the alternative technology using imported combustion engines and fossil fuels. With the high yields obtained from improved rice varieties, the marginal return in potential energy from grain greatly exceeds the energy consumed in the pumping process. (It has been estimated that 30 kg of rice, converted into human energy, will lift enough water to produce nearly 900 kg of rice in Bangladesh.)[17]

POWER PUMPING

To lift 1 m^3/hr through a height of 1 metre is generally estimated to require a power of 6·44 watts[18] or 1/116 hp. The proportionality of these volume and depth relationships is, however, only approximate, and engineers can design the optimum type of pump for each depth and volume. Measured crudely per m^3/hr and metre depth, requirements may vary from 1/130 to 1/240 hp. If working at full efficiency, such a pump will use only 0·0064 kwh/m^3/m depth.[19] In practice, though, much lower efficiencies may be obtained. For a depth of 7·5 m, power consumption was 0·1 kwh/m^3, and for a depth of 3 m, 0·085—more than twice the expected amounts. The Pakistan price was 2·2 c_{74}/kwh, much above US or European prices. Diesel fuel and lubricant were also at about twice US prices. Ghulam Mohammad put diesel-fuel costs per unit output about 20 per cent above electricity costs. At Ghulam Mohammad's prices for fuel and lubricants for a depth of 10 metres, for pumps delivering 100 m^3/hr or more, costs should amount to 0·33 c_{74}/m^3

Finney[20] allows 10 per cent depreciation but only 5 per cent interest, and the latter figure has been raised to 10 per cent, as shown in table 5.3. Finney correctly pointed out, further, that 35–40 per cent of the cost of fuel and lubricant was tax, and that the 'social costs' were considerably lower.

Table 5.3. Costs of irrigation in Pakistan.

	Costs in Pakistan c_{74}/m^3	
	Cotton area	*Rice area (higher water table)*
Interest and depreciation	0·271	0·212
Labour	0·045	0·045
Fuel and lubricant	0·414	0·203
Repair and maintenance	0·068	0·057
	0·798	0·517

Ghulam Mohammad's costs for Pakistan[21] (reckoning interest and depreciation at 10 per cent each) for a working year of 2200 hours are shown in table 5.4.

Table 5.4. Pumping costs in Pakistan.

Costs c_{74}/m^3	*Diesel (102 m^3/hr)*	*Electric (127 m^3/hr)*
Interest and depreciation	0·34	0·16
Fuel and lubricant	0·19	0·23
Labour (including maintenance)	0·11	0·06
	0·64	0·45

Ghulam Mohammad revised World Bank estimates and also made some calculations at 'shadow prices'. This device attempts to take account of social rather than private costs. Allowing for Pakistan's difficulty in securing sufficient export earnings, it may

Table 5.5. Economic costs of tubewell water in Pakistan.

	Government			*Private: electric power*			*Private: diesel*		
Capacity (m^3/hr)	*408*			*102*			*102*		
Extimated life (yrs)	*20*			*10*			*10*		
Valuation of imported goods ratio to face value	1	$1\frac{1}{2}$	2	1	$1\frac{1}{2}$	2	1	$1\frac{1}{2}$	2
Capital cost (000 R)	140	173	243	7	7·7	8·4	9	9·9	10·8
Annual costs (000 R/yr)									
Interest	14·0	17·3	24·3	0·70	0·77	0·84	0·90	0·99	1·08
Depreciation	7·0	8·65	12·15	0·70	0·77	0·84	0·90	0·99	1·08
Operation and maintenance	3·0			0·4			0·8	0·5	0·5
Fuel	8·65			1·9	2·6	3·1	2·5	1·75	2·1
	32·65	37·6	48·1	3·7	4·54	5·18	5·1	4·23	4·76
Water pumped (000 m^3/yr)	1445			247			247		
Water delivered to field	1087			223			223		
Cost R/000 m^3 delivered to field	30·1	34·6	44·2	16·6	20·4	23·2	22·8	19·0	21·3
Do. c_{74}/m^3 (0·25 \$/R)	1·31	1·50	1·94	0·72	0·89	1·02	1·00	0·82	0·93

Source: Ghulam Mohammad, 'Private tubewell development', *op. cit.*

be considered that purchases of imported goods should be debited at 1¼ times or even twice their face value. But by the same token, however, the tax element included in the price of diesel oil should be eliminated in these calculations (see table 5.5). Finally, all interest rates have been expressed at 10 per cent and depreciation calculated at straight line rates.

An official Tunisian report[22] gives data for six schemes requiring pumping, whose average is shown in table 5.6 (conversion is assumed at $ × 0·52 dinars and 10 per cent interest).

Table 5.6. Tunisian pumping costs (c^{74}/m^3).

Interest	3·27
Depreciation	1·24
Fuel and lubricant	1·14
Maintenance and other expenses	0·61
	6·26

Nebbia[23] makes some international comparisons of sub-soil water pumping costs which show economies of scale with size of pump and increasing costs at depth (see table 5.7).

Table 5.7. International comparison of pumping costs at various depths (costs in c_{74}/m^3).

		Depth (m)		
		45	100	
(1) USA (California) Pump 72 m³/hr		1·4	2·9	
288 m³/hr		0·8	1·7	
		Depth (m)		
	50	100	250	350
(2) Italy pump 72 m³/hr Utilization 1500 hrs/yr	2·3	4·5	11·6	15·2
3000 hrs/yr	1·6	2·9	7·1	9·6

(3) Chile, pumping 360 m³/hr, depth 20 m, 0·47 c_{74}/m^3 (including interest at 8 per cent, depreciation at 4 per cent, maintenance at 2 per cent).

Sirohi and Pine[24] suggested a simple formula for irrigating sorghum, of 1·24 c_{74}/m^3 variable costs and 39 $\$_{74}$/ha fixed costs (about 0·78 c_{74}/m^3 if average irrigation water input is 50 cm).

From the 1890s[25] to 1910, the cost of lifting 1 m^3 of water through 10 metres (presumably by steam power) in Arizona was 1·2 c_{74} (of which 0·3 were capital costs). In recent years the cost has fallen to 0·144 c_{74}. The average lift of water in Arizona in metres has been as follows:

1915	9	1948	60	1961	120
1935	30	1955	90	1967	150

In Nebraska[26] a 1953 estimate indicated that with modern equipment working 1000 hrs/yr or more, water could be pumped from 30 metres at 0·98 c_{74}/m^3, but at about twice that cost with older equipment. Costs in modern Italy[27] are similar. However, pumping from shallow wells in the sub-soil at Carnarvon in Western Australia is estimated[28] to cost 1·94 c_{74}/m^3.

Very small projects cannot work at high efficiency. In Vietnam[29] small 5-hp impeller pumps (constructed by local mechanics, with imported engines) lifting 30 m^3/hr from ditches are hired out at about 1·86 c_{74}/m^3. A plan for a small diesel pump to water vegetables for 30–60 African families is costed as high as 12 c_{74}/m^3.[30] On the other hand, Ghulam Mohammad (private tubewell development. *op. cit.*) points out that there may be substantial diseconomies of scale for very large projects pumping 400 m^3/hr. Unit costs of well, power line, and drainage may all rise.

It will be a matter of great interest, however, for India and Pakistan if new low fuel consumption and low capital and maintenance costs are found to be possible for a new type of heat engine which burns oil and vaporizes not water but fluids of high molecular weight, and which has the great advantage that it can be completely sealed, and should need inspection and maintenance only at long intervals. Original work on small turbines using organic working fluids was done by Tabor and Bronicki of the Israeli National Physical Laboratory in connection with the development of a solar power unit.[31] Wilson (Oxford University Department of Engineering Science) and Moss (Esso Research) are trying to develop a cheap and efficient oil-fired version of this machine.

Wynn[32] recorded 101 hectares of vegetable crops in the Sudan irrigated with 870,000 m^3 annually at a lift of 11 metres and a capital cost of $\$_{74}$75,600—that is, 18·7 c_{74}/m^3/yr—and 630 hectares of cotton = *dura* (mixed cultivation) receiving 6,160,000 m^3/yr at a lift of 20 metres, at a capital cost of $\$_{74}$556,000—or 9·0 c_{74}/m^3/yr. However, only 12 per cent of the vegetable cost and 20 per cent of the cotton-*dura* cost represented the cost of the pump and engine (with water supplied by the Nile); all the rest represented the cost of channels, land levelling, and so on—costs which in some other countries had already been met in the remote past.

LABOUR INTENSIVE CONSTRUCTION

In an era with widespread rural unemployment there is considerable intuitive appeal for labour-intensive construction methods. Economists add a theoretical rationale to this by pointing out that in most countries wage rates for unskilled labour are higher than the opportunity cost involved in putting that labour to work. (Conversely foreign exchange is likely to be undervalued.) Wage rates are often set by government as a result of various pressures including trade union bargaining and the net result is to overprice unskilled labour. If labour's opportunity costs or shadow prices (sometimes called accounting prices) are substituted for market prices labour intensive options will look more favourable. In the irrigation field this will favour masonry or earthfill structures rather than concrete.

In Pakistan Carruthers found that the effect of shadow pricing drainage options was to increase the costs of tile drains by 34 per cent but decrease the cost of hand dug open drains by 15 per cent. This changed their ranking but in the Sind Province tubewell drainage was still cheaper than both at nearly half the cost per hectare drained.[33]

Despite the intuitive appeal of shadow pricing there are practical problems involved in undertaking labour-intensive options. First, engineers normally find it difficult to readily obtain information on shadow prices to substitute for market prices in design work. Second, contractors and operating agencies often find it difficult to obtain labour subsidies from government to make up the difference between market prices and shadow prices. You cannot pay labour shadow wages! High minimum wages coupled with over-valued domestic currency, which are typical developing country conditions, are a prime cause of inappropriate technology in irrigation construction. It amounts to a tax on labour and a subsidy upon foreign machinery imports. Third, engineers often contend that manual labour is slower, produces poorer quality work, is more troublesome than machines, and it is often not readily available.

Manual labour is not necessarily slower and for certain tasks it can be cheaper even using market prices. In India the Nagarjuna Sagar Dam was a labour-intensive masonry structure that took 10 years to complete. The Bhakra Dam is concrete and therefore essentially mechanically constructed. This took only seven years to complete but this is a smaller structure and rate of work was equivalent. The International Labour Organization[34] quote Chinese experience which is not necessarily reproducible elsewhere, where in 80 days a main irrigation canal 168 kms long was completed without any machinery. In comparing mechanical and labour intensive options the waiting time for machinery to arrive on site and its actual, as opposed to potential, performance under field conditions has to be assessed.

ILO (*op. cit*) report that for digging and stone crushing manual methods were cheaper on the Sharavathi Valley project in India (for digging hand labour was costing only 25 per cent of mechanical shovel). However, costs were similar for vertical lifting of stones and masonry. They quote an FAO/SIDA source which found that manual lifting of stones beyond 12 metres was uneconomic compared to locally fabricated small cranes.

The quality of work argument against manual labour is hard to sustain except in concrete placement. Quality is a function of management. The troublesome nature of labour is a more plausible reason for choosing machines which may break down but they do not go on strike or leave without notice. On large contracts the numbers of labour required would produce awe-inspiring management problems. On the Hirakud Dam in India ILO (*op. cit*) report that 3 million labourers would have been required to complete it in the time actually taken. Even in India this would strain the management and labour resources.

Unemployment in rural areas is often more apparent than real. In India, Aswan in Egypt and even in China (Kwanting Reservoir) labour has had to be imported long distances, often to the great surprise of the engineers. Agricultural labour availability is always likely to be seasonal and therefore hard to recruit at harvest or sowing times. For example, ILO (*op. cit*). report that on the Kandana Project in Gujarat, India, daily wages were increased from Rs.3·5 to Rs.10·0 at peak periods and during festivals. Very seldom is the pace of irrigation construction in line with the agricultural calendar. Healey found that comparing the Damooder Valley Project, Bhakra Nangal Project and Hirakud Project, only on the last did the construction activity dovetail with agriculture.[35]

The labour intensive technologies appear more suitable and more easily managed when used for canals and drains which, like roads, are spread through the countryside and not concentrated for long periods in one constrained site.

INCREASING MARGINAL COSTS

Simple diversion from a flowing stream can be very cheap indeed. The need for damming increases the cost. Irrigation from wells, where sub-soil water is available, is more costly. It involves the initial capital cost of digging the well and installing a pump, and in some cases of supplying a power line to the pump. Both capital cost and operating cost primarily depend upon the depth to which the well has to be sunk.

Naturally the first sites to be developed in any country are the most favoured, with low cost per hectare irrigated. In the United States there were many simple 'run-of-river' diversions to adjacent flood plains, and in some cases not much irrigation was required because of reliable rainfall. The increasing marginal costs are shown strikingly in figure 5.1.

In the United States in 1939, the average costs of maintenance and operation, expressed in money of 1974 purchasing power, were 0·28 c_{74}/m^3 delivered, or $\$_{74}$/24·1/ha irrigated, of which 70 per cent were interest and redemption charges on indebtedness. The average depth of water delivered was 0·85 m, to obtain which result, however, an average of 1·37 m had to enter the canals, nearly 40 per cent being lost by seepage and evaporation.

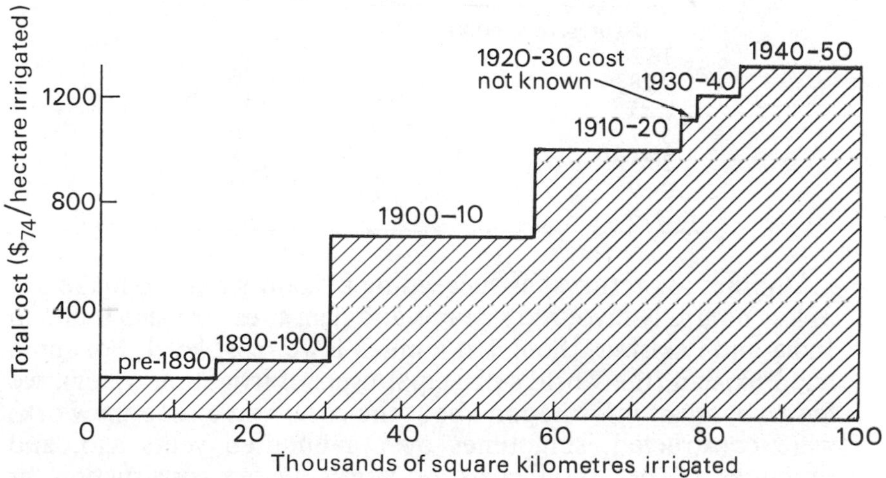

Figure 5.1. Irrigation costs in the USA.

Per unit delivered, the cost was 0·19 c_{74}/m^3 with water drawn by gravity from the natural flow of streams, 0·57 with water pumped from streams, and 1·19 for water pumped from wells. In real terms, these costs were much the same over the whole period 1909–49. The average lift in pumping water from wells in the United States was 17 m (rather more than in India).

An interesting projection for costs from succeeding new dams which might be built in the Upper Missouri region shows rising marginal costs (see table 5.8).[36] But at the other end of the cost scale, in the Rio Grande-Pecos region, where present storage is 7·3 billion m^3, it is proposed to raise it to 9·1 billion m^3. However, the marginal cost for this, at 1974 levels of purchasing power, is 61 c_{74}/m^3. At this rate, the committee points out, all irrigated agriculture in this region will have to be abandoned—perhaps some urban settlements too.

A similar situation, but with very much higher costs, is shown on the Great Ouse River in England.[37] To raise the present disposable amount of 1 million m^3/day to 1·5 million, by using low-level groundwater and pumped storage, will show a marginal cost of 16 c_{74/m^3}. A barrage in the Wash at the mouth of the river would create a wide but shallow fresh water lake, raising disposable supplies to 2·7 million/m^3/day, but at a marginal cost of 22 c_{74}/m^3.

Table 5.8. Marginal costs of water from the Upper Missouri.

Dependable flow ($m^3 \times 10^{10}$/yr)	*Marginal costs (c_{74}/m^3) for additional supply*
0·25 (present supply)	
1·73	0·10
2·28	0·26
2·83	0·31
3·17	0·40
3·45	0·61
3·59	0·70
3·73	0·86

Diminishing returns, as succeeding irrigators have to make use of less naturally favoured streams and dam sites, are also found in India and Pakistan, though at a much lower cost level. For India and Pakistan (the world's principal users of irrigation water), we have abundant information about the costs at the time the works were constructed, sometimes over a hundred years ago, and estimates of the total costs of works under construction or investigation now.[38] All costs are again expressed in standard

Table 5.9. Dam irrigation in India and Pakistan (excluding Burma).

Date	Total area irrigated (million ha)	Specified irrigation works: Period of construction	Area irrigated (million ha)	Average cost ($\$_{74}$ per ha irrigated)
		Before 1900	5·1	199
1901	12·5	1901–10	1·1	152
1911	17	1911–20	1·3	219
1921	18	1921–30	0·75	267
1931	19	1931–40	1·2	376

unit, namely a dollar of 1974 purchasing power (see table 5.9). The conversion of past figures to present-day costs is performed by means of index numbers whose precision is not great; however, we require only orders of magnitude. In India and Pakistan, most of the growth of the cultivated area over this period has been through irrigation.

Some indication of more recent capital costs for projects in India can be obtained from a World Bank evaluation.[39] The capital costs of three projects completed during the 1960s are shown in table 5.10.

Table 5.10. Capital costs of major irrigation schemes in India.

	Completed	Cost ($\$_{74}$/ha)	Internal rate of return (%)
Shetoniji	1964	820	17·4
Salandhi	1965	784	17·3
Purna	1966	809	11·2

Unit costs in India are clearly still rising, being double the cost of schemes constructed during the 1930s (in terms of 1974 purchasing power).

As less favourable sites are selected, increases in capital cost can be offset by technological developments. In the last 15 years major developments in the design of concrete dams, the design and operation of earth- and rock-moving equipment, in drilling techniques, and the use of cheap plastic piping have all lowered the potential cost of any one site. A US Senate Select Committee wrote: ‘Investigation of historical data revealed no clearly established trend over time in costs per unit of capacity.’ The expected rise in real costs due to having used up the best sites first was roughly offset by declines which resulted from technological advances in dam construction. The Select Committee appear to be referring to the period since about 1940; previously, as shown in fig 5.1, real costs were rising.

RANGE OF CAPITAL COSTS

Capital cost per hectare developed on per m^3 of capacity is a piece of information sometimes used to compare alternative schemes. Before presenting such information, it is worthwhile reminding the reader that in irrigation engineering there is considerable scope for substituting capital by recurrent costs. For example, low capital cost structures will tend to generate high recurrent costs and/or high replacement costs. Tubewells can be designed for low running costs by selecting wells of larger diameter or greater depth. Gravity supply systems tend to have high capital costs but low recurrent costs compared to pump supply alternatives. The optimal mix of capital and recurrent cost which gives the least total cost solution can be determined only in relation to the endowment of local resources and expectations on shifts in availability of those resources over a period of time. To a large extent, the discount rate selected for comparing alternatives will be crucial in determining the optimal mix of capital and recurrent costs. A high discount rate will favour increased emphasis upon capital saving by increased recurrent costs. For these reasons a comparison of schemes based on capital costs gives an extremely partial insight.

The best and most recent aggregate estimates of irrigation costs for new extensions that are available were prepared for the 1974 UN World Food Conference. They estimated costs ranging from \$1500 per hectare in Latin America and the Far East to nearly \$2500 per hectare in the Near East and Africa. Rehabilitation of existing under-utilized schemes was only 20–30 per cent of the cost of these new extensions. The Ganges-Kabadak Project in Bangladesh is an irrigation and flood protection scheme. If the present value of estimated flood protection benefits (60 years at 10 per cent discount rate) are subtracted from capital costs at 1974 values the net costs for irrigation are \$1420 per hectare.[40] Similar cost was found in Turkey. The first stage of a large (170,000-ha) irrigation project in south-east Turkey, constructed in the mid-1960s, cost \$1480/ha (Otten and Reutlinger, note 39). Estimates of the costs of irrigation developments over a ten-year period in various parts of the world are given in table 5.11 below.

Rehabilitation is a neglected activity. The World Food Con-

Table 5.11. Estimated costs of irrigation developments, 1975–85 ($\$_{74}$/ha).

Area	*Renovation & improvement of existing irrigated areas*	*New irrigation areas*
Far East	418	1466
Near East	560	2467
Africa	500	2400
Latin America	420	1540

Source: Secretariat Proposals. UN World Food Conference, Rome, 1974.

ference estimated costs at two-thirds that of new schemes for a given incremental output. The main problems are that the potential productivity is often unknown to the operators and the effects of neglect may not be evident. Even where the efficiency of an irrigation system is evidently low, it is often impossible to execute worthwhile renovation because of a lack of skilled manpower and financial resources.

Rehabilitation projects are becoming a fashionable form of aid. This is not simply because they are extremely economic investments as a consequence of the large element of sunk costs, but also because of the politics of aid. Few donor governments allow aid for recurrent budget support. However, extreme shortages of recurrent financial resources are often the main reason why rehabilitation becomes necessary. Therefore donor-aided rehabilitation projects become, in effect, an extremely inefficient form of recurrent budget support.

It is difficult to support the assertion that up-grading of existing irrigation networks deserve first priority. Certainly it is possible to identify a very wide range of yields and net returns on different schemes. The existence of a huge, unexploited potential is well established. However, there are doubts on the managerial ability of operating agencies and their will to tackle the constraining and intransigent problems of an agricultural and institutional nature.

An economic evaluation of irrigation projects in Mexico, conducted by the World Bank,[41] illustrates some of the difficulties that are generated by apparently profitable rehabilitation projects. The major project was for 237,000 hectares in north-western Mexico, one-third of which was saline, the remainder having reduced yields. The costs over the period 1962–6 were only $\$_{74}$180 per hectare, which included new and improved drains, canals and water control structures, road improvement, telephones, dwellings for the operation staff, and construction equipment. This has been profitable, yielding an estimated rate of return which varied for sub-projects from 8 to 18 per cent. However, the full potential was not exploited. Many of the necessary on-farm works did not materialize, and only half the saline land that was planned to be reclaimed was brought back into cultivation. Much of the benefit consisted not of reclamation, but of preventing land from deteriorating further; of course, the benefits are no less real for that. A second project in north-eastern Mexico gave lower returns, mainly because the necessary on-farm private investments in complementary facilities were not undertaken.

Rehabilitation is no guarantee that a project will thereafter perform well, Mitra reports that reclaimed and developed farming units on the Greater Mussayib Project Area in Iraq started deteriorating to traditional inefficiency immediately after special rehabilitation project operations were withdrawn.[42] He estimates that within ten years the cropping intensity drops from 115 per cent back to 45 per cent. He blames this deterioration on the failure of government to maintain irrigation and drainage networks, to provide adequate agricultural extension service, and to supply an efficient farm credit system. The cost of this temporary increase in productivity is nearly $160 per hectare.

Reliable measures to effect rehabilitation are often difficult to prescribe. Clayton *et al.*[43] evaluated a groundwater project in Jordan, and found that poor maintenance of the diesel generators threatened the whole project. Within the existing institutional framework there appeared to be no obvious remedy. Declarations such as 'manufacturers' recommended lubricating procedures must be followed' are clearly inadequate. Similarly, knowledge of how, precisely, farmers are to improve the efficient use of irrigation water is simpler to debate than to implement. More emphasis is required than hitherto upon management systems and on-farm micro-studies if the evidently poor performance of many irrigation projects is to be improved.

Rehabilitation of existing irrigation is often a pre-condition for other necessary improvements. Shoaib Sultan Khan[44] describes one of the most instructive experiments in 'integrated rural development' (a revamped 1970s phrase for the ill-fated Community Development). It describes his attempts to change the still-extant British colonial system of local government, which emphasized control and revenue functions, to a development-orientated system. A survey of needs at the village level indicated that land and water rehabilitation would have to precede direct production programmes. Of 14,000 hectares of land in the Daudzai area, nearly 12,000 hectares were damaged by waterlogging or salinity, floods or scouring by the river. Furthermore, the irrigation system was falling apart, and the situation was deteriorating. There is clearly little hope of production programmes until and unless the land and water resources are safeguarded. Villagers were well aware of this, and some, for example, refused improved seeds, fertilizers, and so on, until shifting sand dunes which regularly blocked their irrigation canal were stabilized. There are virtually no costings in the report, but it is clear from the description that rehabilitation measures were extremely cost-effective; particularly so, if benefits from later use of modern agricultural inputs is partly attributed to rehabilitation.

In the first chapter, the authors deplored the dearth of *ex-post* evaluations which fulfil the function of providing feedback to planners. However, such evaluations also have the important direct function of indicating desirable changes in the execution of existing projects. Rehabilitation of existing works which were normally constructed in the most favoured situations is generally a more cost-effective use for public resources than extension of irrigation into new areas. Such rehabilitation might cost from one-third to one-fifth as much as new projects and might yield half the net benefit. However, it is accepted that many more practical evaluations of on-going schemes and rehabilitation projects are required to verify this proposition. The UN target of 46 million hectares to be renovated between 1975 and 1985 at a cost of $21,000 million is desirable but optimistic.

OPERATION AND MAINTENANCE COSTS

If operation and maintenance was carried out efficiently there would be no need for rehabilitation. Operation and maintenance standards are generally reported to be poor. There are several possible reasons for this. Recurrent finance is the most scarce resource and it cannot generally be obtained as aid. Indeed large capital aid projects can exacerbate the recurrent finance allocation problems. Lack of political will to obtain revenue or lack of repayment capacity if farmers are poor, reported in chapter 7, can also add to difficulties. Ancillary facilities such as field channels, water course diversion structures, and field drains are often incomplete. Discipline among farmers is often lax and those near the head may make illegal offtakes and withdraw too much water. Operation and maintenance, once starved of funds and equipment, becomes a low prestige area and has difficulty in attracting and retaining high calibre staff.

Operation and maintenance costs can be substantial. Ellman and Pingle[45] estimate that annually $\$_{78}$6·3 to 10·1 per hectare are required for large-scale schemes in India (although only $\$_{78}$3·1 is provided). For private and public tubewells costs are $\$_{78}$25–44 per hectare. In Bangladesh they claim surface schemes obtain $\$_{78}$8·4 per hectare and low-lift pumps $\$_{78}$45·8 per hectares. This is the finance that is supplied but the poor standard of maintenance suggests that this is inadequate.

In the Indian or Bangladesh surface irrigation case the ratio of recurrent to capital costs is close to 1 to 70–100 so that it is understandable if the recurrent element is discounted. (In project appraisal it is literally discounted, normally at high, 10–15 per cent rates of discount.) However, the deteriorated state of many contemporary modern irrigation projects indicates how short-sighted such a policy can be.

The experience of the greater Mussayib project in Iraq (which had few if any real financial problems) should be a warning to all irrigation authorities:

> Initially, the Government made efforts to develop the Greater Mussayib area during the period 1953–6. The first settlers took up lands in 1956 but a considerable number of farming units were taken up by absentee landlords and actual farming was performed by tenant farmers. The project was supervized by a small government administrative unit with limited powers and facilities. No services were provided to the farmers who continued to follow traditional farm management practices. With neither the funds nor equipment to undertake maintenance of canals and drains, the supervising government unit was soon confronted with rapidly silting canals and branches, overgrown and clogged drains and increasing salinity which gradually forces the farmers to adopt extensive farming practices. Productivity soon declined and yields of grains dropped to the lowest level for irrigated farming is the Near East region.[46]

THE ECONOMICS OF STORAGE

Storage enables water that is held from periods of low value or surplus to be used in periods of scarcity. In many arid areas there are periods of high river flows stemming from local rainfall or, more usually, rainfall or snow melt in the catchment. The Nile flood is perhaps the best-known example of peaking of the river flows. This flood is caused by rainfall in Ethiopia. In the Indus system, two-thirds of the discharge passes unused to the sea in the summer months. The main source is snow melt in the Hindu Kush and Himalayan ranges, supplemented by monsoon rainfall in north-western India. Because all the winter water is now used, further irrigation development is dependent upon seasonal (summer only) canals, groundwater, or seasonal canals made perennial by release of stored water.

In establishing the economics of storage schemes, the cost per m^3 is a useful but insufficient piece of information. The supply of water over the life of the dam can be expected to fall as siltation removes storage capacity. In establishing the life of storage capacity, the concept of half-life, as used in atomic physics, is useful. The half-life of a dam is the number of years before 50 per cent of the storage potential is removed by siltation. Other valuable concepts are live storage and super-storage. The former is that proportion of storage capacity available in normal use (the balance is dead storage); the latter is the safe storage available for short periods. Sometimes it is used to retain a temporary flood peak.

Dams vary widely in the anticipated life of their storage. The Warsak Dam on the Kabul River had a half-life of less than ten years. The Indus River is depositing 400 million tons of silt per year in the Tarbela Reservoir. Within 50 years the reservoir will be completely filled, and there will only be a residual storage equivalent ot 10–15 per cent of the initial level. The Mangla Dam on the nearby Jhelum River will lose only 30 per cent of its storage over the same period.

A heavy silt load affects the economics of dams by causing abrasion to structures and turbines, and it also increases the operating cost of irrigation canals. In some instances, silting could be reduced. For example, much of the silt load to the Mangla Dam stems from poor catchment, conservation, and erosion control. Conservation, though technically feasible, is unlikely to be undertaken because the benefits, when discounted to present value terms, compare unfavourably with alternatives. Furthermore, there is another complication in the fact that the source of the silt is largely in India, whereas the benefits would be realized in Pakistan. Other examples of the same problem can be found, where the potential benefits of an activity in which an individual, project, or country incurs the cost but does not gain the benefits. For example, the Tarbela Dam acts as a silt trap which, for as long as it is operational, will protect any downstream storage works. Therefore, given down-stream development, the apparent cost of Tarbela water overstates the potential economic cost.

There are other means which can be used to prolong storage life and thereby to lower the unit costs of development. If site conditions permit, sluices can be installed to flush silt down-stream. Alternatively, off-stream storage can be built, to be filled by diversion from the river when the hydrograph is falling and the silt load is markedly lower than at rising or peak discharge. This system is planned for the Sehwan-Manchar complex on the Lower Indus, which will store $1{\cdot}8\ m^3 \times 10^9$ of irrigation water for winter use.

It is a sound general principle to use stored water as soon as possible after river flow subsides. In this way the total volume of water available is maximized, because seepage and evaporation losses are minimized. However, it is always worth considering whether it might not be worthwhile to accept storage losses by retaining the water for a longer period when it might have a higher total value.

This principle can be illustrated with reference to design decisions on the Lower Indus Project. In this project, it is planned to store nearly $5{\cdot}0\ m^3 \times 10^9$ of water and use it in the four months immediately after the reservoirs are full—namely, November to February. With this procedure, $1{\cdot}23\ m^3 \times 10^9$ are lost and the balance delivered to the canals. About half the losses are due to evaporation and half to seepage. If the water were all held to be used in May, the losses would rise to $3{\cdot}08\ m^3 \times 10^9$, 60 per cent of which would be evaporation. The amount of water delivered to the canals with this operating procedure is thus 50 per cent of that with the normal early release pattern. The ratio of average return per unit of water in each period was 1·3:1 in favour of May release; to be an economic proposition, the return in favour of May water had to exceed 2:1. It is, of course, the marginal return which is relevant, but in the Lower Indus cropping intensities are very low, less than 50 per cent, with water the limiting factor, and therefore average and marginal returns are close. In this instance, the normal practice of using stored water immediately after river levels fall is justified.

The costs of storage are very largely determined by site conditions. In broad valleys with poor foundation conditions, costs will be high. An ideal site is a narrow gorge broadening to a large valley up-stream with good, solid rock conditions and no risk of earthquake activity. The Tarbela Dam in Pakistan has poor foundation conditions and a low storage volume, resulting in high-cost water. It is built in a broad shallow valley with a deficient base. An impervious earth blanket was laid 1–5 km up-stream to prevent seepage. In the early months of operation there has been some anxiety on the effectiveness of this blanket.

Storage costs can also be affected by conditions up-stream. About 20 per cent of the construction costs at Mangla were incurred in constructing a flood spillway; it would have been greater, but there is super-storage at Mangla equivalent to nearly 40 per cent of normal storage volume. Flooding on the Jhelum River at Mangla is variable and unpredictable, as it depends upon monsoon rain. In contrast, the Indus at Tarbela has a lower flood peak as it is largely dependent upon snow melt. However, there is less super-storage available, and there is a risk of very large flood peaks from the failure of natural ice dams which occur in the Himalayas. In consequence, despite lower flood peaks there is even greater spillway capacity at Tarbela.

STORAGE COSTS

In the mid-1960s Colin Clark made a study of capital costs of water storage in 24 dams in Australia, Ceylon, Egypt, India, Italy, Mexico, and Pakistan covering a period of more than 50 years. He found that, in real terms, costs varied over a certain (wide) range, that costs per unit fell with scale, but that the expected increase costs for later construction (for example, the initial use of cheap sites) was not obvious. The scale relationships, at 1974 prices are illustrated in fig. 5.2.[47]

A quite inordinately high cost of 43–86 c_{74}/m^3 (£500–1000/ million gallons) is estimated[48] for small dams in England (apparently below 1000 m^3). In some of these very small dams the ratio of the volume of water stored to the volume of earth excavated was not much greater than 1; no wonder the cost was high. (By 1978 this cost had risen to 70–110 c/m^3 and Nix notes that lining, presumably with a butyl sheet, would double this cost.)

The Italian figures show that 5 m^3 of water are stored, on the average, for each cubic metre of dam wall (a figure of 4 is considered to be the lower economic limit); whereas in large dams this ratio rises enormously. The figure stands at 600 in hydro-electric projects in Scotland, 1500 in the Hoover Dam, and 1,000,000 in the Owen Falls Dam in Africa at the outlet of Lake Victoria. In the very large Tarbela Project in Pakistan, the ratio, however, is only 84.

Some interesting comparisons collected by Kanwar Sain,

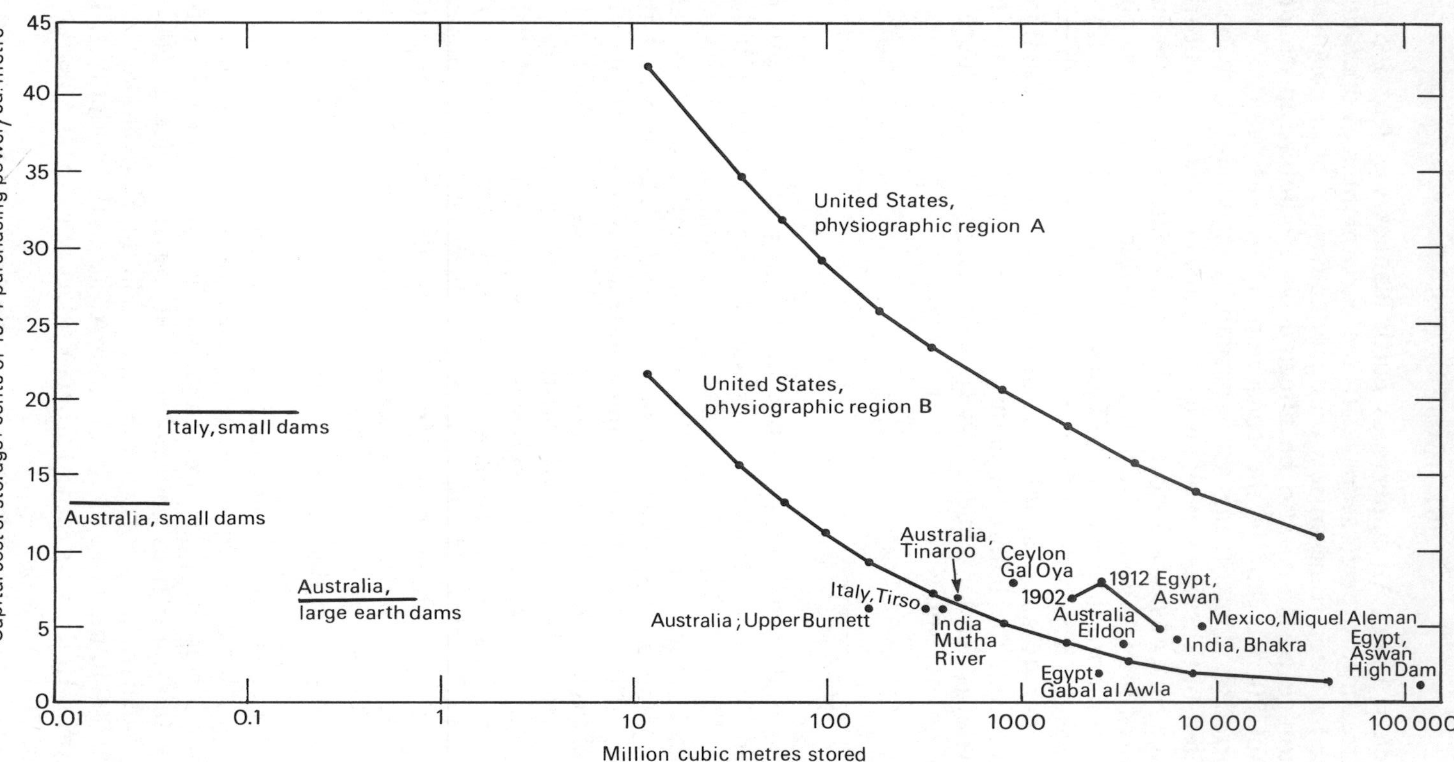

Figure 5.2. Capital costs of water storage.

Director General for Irrigation in India, showed that, in a number of important dams in India and elsewhere, built before the 1930s, when hand labour methods predominated, the cost of the dam, per cubic metre of water stored, generally worked out at approximately 4·5 c_{74} almost irrespective of the size of the dam.[49] In the larger dams, a great deal more water is stored per unit volume of the dam; but the dam structure itself has to be so much more massive and costly.

What is gained by building large areas for water storage is, however, to a considerable extent lost again in the costs of the distribution network and of land levelling which are necessary to accompany them. Thus, in the Australian examples, the ratio of these costs to the cost of storage was 2·15 for Tinaroo, 0·25 for the large earth dams, and nothing for spray irrigation from small dams. (Spray irrigation is comparatively costly in equipment, power, and labour, but by its nature calls for no canals or ground levelling.) In southern Italy, the ratio was estimated at 1·25 for distribution plus 0·37 for levelling, and in Italian hill lakes at about 2·3 for distribution.[50] In India, the ratio was 0·81 for the Mutha River Project, 0·9 for Hirakud, and 2·2 for Bhakra, where, for political reasons, the water had to be distributed over a much wider area than was economically justifiable. The Indus in Pakistan has an immense summer flow from melting snows in the mountains, which is controlled by a comparatively low-cost barrage at Sukkur. In this exceptional case, the cost of canals and water courses was four times that of the barrage and headworks.

There have been some notable examples of failure of irrigation schemes, because the authorities economized or ignored on-farm land development and levelling costs. An apparently low ratio of storage to distribution cost is sometimes illusory. For example, irrigation water was largely unused in the Purna scheme in India because land preparation was left to farmers who expected little or no benefits from it. Similarly, in rehabilitation projects in north-east and north-west Mexico financed by the World Bank, virtually none of the expected private on-farm investment took place. In the case of drainage ditches, where the benefits are less evident and the returns are not immediate, the prospect for voluntary, intensive labour by farmers is even less bright. With the Sukkur Barrage, although distribution costs were four times that of headworks, no drainage was included, which has lead, in time (approximately 20 years) to land salinization, depressed crop yields, and other problems.

In some pumped schemes, temporary diurnal storage may be desirable to avoid night irrigation, to avoid peak daily evaporation periods, or, with sprinklers, to avoid periods of the day when high winds are expected. Similarly, it may be necessary to pump wells for, say, 20 hours per day and store water in a balancing reservoir in order to operate canals evenly for 24 hours per day. In Jordan, the FAO found little difference in cost between concrete reservoirs or butyl-lined earth reservoirs. Concrete is cheaper up to 2000 m^3 and easier to construct.

Official estimates by a US Select Committee[51] suggest that additional water could be obtained by damming the Upper Missouri to the extent of 3·4 million m^3/day at a bulk cost of only 0·7 c_{74}/m^3. But, at the other end of the scale, in the Rio Grand-Pecos region, where present storage is 7·3 thousand million m^3 and it is proposed to raise it to 9·1 thousand million m^3, annual capital charges will rise from $\$_{74}$14·2 million to $\$_{74}$22·5 million, for a not greatly increased flow (14 million m^3/yr) representing a marginal cost of no less than 59 c_{74}/m^3. At this rate, the committee points out, all irrigated agriculture in this region will have to be abandoned—and perhaps some urban settlements too.

It appears that costs in California have been rising very rapidly since the best dam sites were used up. In 1925, a transition period, costs[52] in c_{74}/m^3 were 1·2 for the public utilities and 1·4 for the old, established irrigation districts, and 2·1 for the Sacramento and San Joaquin schemes, but had already risen to 3·0 for some Mutual Water Companies and to 6·0 for the Santa Clare scheme.

In 1974 values, water from the immense new Feather River project in California[53] will cost 11·3 c_{74}/m^3 at 5 per cent interest. At the rate actually paid by the state (2·7 per cent), the cost falls to 7·2 on a total capital cost of $\$_{74}$2·6 thousand million. As always, the estimate rose. In 1960, the electors of California agreed (by a narrow margin) to a bond issue of $1·75 thousand million to cover only a first instalment of the project.[54] The authors contend that the project was premature, that an increase in water rates should have come first, and that cheaper alternatives were available both from reclamation of waste waters and from reallocation of agricultural supplies. It might also be pointed out that throughout California water has been priced below the marginal cost of existing supply, not to mention the high cost from the river supply.

A general estimate[55] for the eastern United States divides capital costs into 30 per cent basic sources, 20 per cent transmission and treatment, and 50 per cent local distribution and storage. Subject to the warning above about rising distribution costs from large dams, it is clear that unit costs fall with size. It is customary in engineering to express costs as a function of size by means of exponents. If the exponent is less than 1, costs rise less rapidly than size—that is, costs per unit fall. (Exponents are best estimated by plotting both size and cost variables logarithmically; if the log of size rises by 1, and the log of cost per unit falls by 0·3, log of total cost rises 0·7.) An exponent as low as 0·56 for the capital cost has been suggested.[56] Adjusting past costs, the authors find, on British evidence:

$$\text{Capital cost in } \$_{74} = 0{\cdot}0163\ (\text{m}^3/\text{day yield})\ 0{\cdot}56$$

The world data reviewed suggest an exponent of about 0·7—in other words, a tenfold increase in size will reduce costs per m^3 by a factor of 2.

THERMAL POWER VERSUS HYDRO-POWER

During the 1960s availability of low-cost fossil fuels and optimism on the prospects for development of nuclear energy resulted in a virtual elimination of new hydro-power installations. Those hydro projects which were initiated during that period were mostly multi-purpose projects; very often irrigation development was their prime function. However, in the 1970s, rapidly escalating fossil-fuel prices, doubts on the environmental impact of nuclear plants, and previous costing of nuclear energy have reviewed interest in hydro as a relatively clean source of energy.[57] At the same time, the energy requirements of new agricultural technology have increased markedly. The shifts in energy demand are both indirect (for example, energy for fertilizer manufacture) and direct (power, for instance, for tubewell pumping).

Often hydro-power development and irrigation are complementary. Water retained by dams for power generation in the dry season can be used down-stream for irrigation. Sometimes the storage of water for hydro-power is a pre-condition to irrigation development. For example, the 50,000 hectares of irrigable land in the Lower Tana of Kenya cannot profitably be irrigated until up-stream storage is completed. This storage is required primarily for hydro-power.

Whether irrigation water is to be used up-stream or down-stream of the hydro facility, there may be a conflict between the water demands of agriculture and power. Water requirements for crops in dry seasons may exceed the average demand by a factor of 2 to 4; power requirements at this time also tend to increase, but by a much lower factor. It may therefore be necessary to increase storage capacity simply to cope with peak irrigation requirements, or to make a choice between releases to meet peak irrigation demand or the withholding of this water for power use in the dry season.

There may be further conflict if flood protection is another function for the reservoir. If a reservoir is a flood retention basin, it has to be kept as near empty as possible. If it is for irrigation or power, it should be kept as full as possible. The extent of conflict depends very largely on the accuracy of flood forecasts. If floods occur only in a well-defined rainy season, it may be possible to operate a reservoir both for flood retention and storage functions. Normally, flood irrigation is comparatively a less important function. The Hirakud Dam in Orissa, an area subject to floods, was estimated[58] to have a value for flood-control purposes of only 1·8 million rupees/yr, which, when capitalized, represents less than 3 per cent of its gross cost.

The fact that in a multi-purpose project water can be used either for irrigation or power generation can be an advantage. The Purna project in India, seven years after completion, had achieved only 15 per cent of projected irrigated area. The water requirements of crops had been over-estimated, agricultural labour was locally scarce, yields under irrigation were not reaching forecast levels, and necessary supporting off-farm and on-farm infrastructure had not been executed.[59] Consequently, there was excess irrigation water available. Extra power-generating capacity was installed, and by 1970–1 actual firm and secondary power generation was 2·4 times the forecast level. It is clear that in future, if the irrigation demand increases, very difficult decisions will need to be taken on the trade-off between hydro and irrigation water. However, given shifts in construction costs for coal-burning power plants ($>\$_{74}125$/kw), physical coal shortages and the large foreign exchange component of nuclear power stations, we can expect more, not less, emphasis on hydro-power production.

ALLOCATING JOINT COSTS

Multi-purpose projects raise important issues of principle regarding economic and financial criteria. In the economic analysis, over-all return has to be satisfactory, and, in addition, incremental benefits have to exceed incremental costs, for each function of the project, by the defined margin. In the financial analysis, there are a number of procedures for allocation of joint costs to the various project purposes. The general principles are that no activity should be assigned costs greater than the value of benefits generated by that activity. At the same time, all the costs that are easily identified as generated by that activity should be allocated to that activity. These are 'separable costs', and represent the minimum that should be charged for that purpose. For example, in a multi-purpose dam scheme, canal operating costs are clearly separable and are the minimum charge to irrigation. Irrigation must bear these costs or be deleted from the project. However, when alternative means exist for providing the same amount of irrigation, for instance, by tubewells, then the cost of providing water by an alternative means provides a basis for charging in a multi-purpose project. The maximum charge for any activity in a multi-purpose project can be represented by the cost of providing an equivalent service by a single-purpose alternative scheme.

The procedure for joint cost allocation[60] requires consideration of all capital and operating costs and benefits in discounted terms at present value. In the first stage, total costs and the benefits to each separate purpose are established. Alternative costs for generating each benefit are then established, and the lower of either benefit or alternative cost is termed justifiable expenditure for each purpose. Separable costs are deleted from justifiable expenditure, and the percentage of the total remaining justifiable expenditure in each purpose is calculated. The total remaining joint cost is then allocated to each purpose according to this percentage distribution. The total allocated joint costs can then be determined. This indicates what would be a reasonable financial charge. As will be shown in chapter 8, before decisions can be taken on what precisely should be paid by beneficiaries of the service, social, political, legal, and other issues must be considered.

DISTRIBUTION COSTS

Irrigation water may be distributed to fields by unlined or lined channels and to crops by furrows or flood basins. Alternatively, it may be carried in pipelines and applied by sprinklers or the newer trickle systems. In Israel, capital costs, set out in table 5.12, increase 30 times from the simplest systems to the most sophisticated. However, with trickle systems, labour requirement falls by nearly the same proportion and field efficiency of water use almost doubles. In an economy where labour is a relatively scarce and expensive factor of production, and where total water supplies are limited, the expensive trickle systems become economic.

Table 5.12. Capital investment, labour requirement, and field efficiency in Israel.

Means of water distribution	Irrigation system Capital investment $\$_{74}$/ha	Ha/man day	Field Efficiency (%)
Surface: Earth channels	6·3	1·5	50
Concrete channels	31·6	2·2	70
Sprinklers: Standard portable equipment	37·9	2·6	75
Semi-permanent hand drag	44·3	3·6	80
Tractor towed	56·9	6·1	75
Permanent solid set	107·5	40·5	85
Sprinklers with remote control	126·5	60·7	90
Trickle: Orchard systems	82·2	40·5	90
Orchard systems with remote control	101·2	60·7	95
Vegetable systems	183·3	40·5	90
Vegetable systems with remote control	202·3	60·7	95

Source: Quoted by J. Ingram in *Report of Study Tour* published by UK Farming Scholarships Trust, 1975.

With a capital investment of $\$_{74}$100 to 200 per hectare, twice the area can be irrigated with a given quantity of water and the man-power requirement is cut drastically.

In Jordan, the FAO compared various conveyance systems and irrigation methods (see table 5.13 below). Earth canals were

Table 5.13. Comparison of costs for sprinkler buried pipe/basin systems in Jordan (Jordan dinar converted at $3·25).

	Field crops		Orchards	
	Basin	Sprinklers	Basin	Sprinklers
Capital ($\$_{74}$/ha)	1360	1360	1170	1463
Total Annual costs ($\$_{74}$/ha/yr)	67·6	60·5	28·9	34·5
Labour (man hours/ha/yr)	180	100	520	70

Source: FAO Medium Term Plan for Development of Groundwater Irrigation in East Jordan, Amman, 1974.

ruled out because of water losses and management problems. With low discharges involved (<1 m^3/sec), buried asbestos cement pipes were a similar cost to concrete-lined channels, but the annual costs of the latter were higher. Water could be applied at the field through border strips, levelled basins, or sprinklers. Border strips had slightly lower annual costs than basins, but higher skills are required. Basins were preferred by the planners. Sprinklers had lower labour requirements than basins (70–100 man-hours/yr compared to 180–520 man-hours/yr for basins) but higher direct costs. However, if the indirect costs and other advantages of sprinklers are considered, sprinklers are cheaper for field crops. For orchard use, basins were preferred. The advantages of sprinklers which have to be considered in comparing costs include a higher proportion of the land under crops (fewer ditches), less levelling, higher application efficiency, higher efficiency of farm machinery, and elimination of night work.

In Australia, Waring[61] had estimated 2·13 c_{74}/m^3 as the cost of irrigating pastures, which, however, he assessed as being above the economic limit, on the grounds that the same gross marginal yield of butterfat could have been obtained by spending 1·75 c_{74} on purchasing 0·17 kg starch equivalent in the form of sorghum. This latter calculation, however, was based on the assumption of a marginal return of 0·2 kg butterfat/kg starch equivalent. Experiments have shown, though, that even the best cows have a marginal return to feeding with grains of only about one-third of this. At Australian prices of butterfat the economic limit to the cost of irrigation water might be as high as 5 c_{74}/m^3.

Australian irrigation has an unfortunate record. Works on the River Murray[62] (quite apart from capital costs in the irrigation areas) have cost 11·6 $c_{74}/m^3/yr$ made available for irrigation. The Mareeba-Dimbulah Scheme in Queensland[63] showed a record capital cost of over $\$1{\cdot}55_{74}/m^3$ stored and 9 c_{74}/m^3 delivered. A small plan[64] to irrigate 70 hectares (that is, enough to keep one man fully occupied) in western Victoria, using both dam and well water, was costed at 2·6 c_{74}/m^3—quite unremunerative as a private venture.

In Italy, the costs of the distribution networks,[65] assuming average irrigation inputs of two-thirds of a metre from flowing streams in Venetia were estimated at 10·7 $c_{74}/m^3/yr$; in the hill lake country, 9·0; and in southern Italy (with a more difficult topography) 17·8 for distribution and 5·3 for levelling.

Table 5.14 gives estimates of total cost in c_{74}/m^3 of irrigation water delivered (sources as quoted above for capital costs, unless otherwise indicated).

Table 5.14. General review of costs (c_{74}/m^3).

Australia	Waring (*op. cit.*) spray irrigation	2·8–5·2[a]
Australia	A. B. Ritchie,[1] Penshurst, Victoria	3·9[b]
Australia	Mareeba-Dimbulah Scheme[2]	1·1[c]
Iran	Oanats[3]	0·28[d]
Italy	Venetia, from streams	0·5[e]
	Central Italy	5·2
	Southern Italy	3·8[f]
	Waste water recovery[4]	3·3
	Spray irrigation, plains	4·4
	Spray irrigation, hilly country[5]	15·7
United Kingdom	Sewage effluent[6]	3·5[g]
	Farm irrigation[7] including costs of well or dam	19·7
	Farm irrigation from streams, below 10,000 m³	20·5
	Farm irrigation from streams, 10,000–14,000 m³	11·2

Sources:

1. A. B. Ritchie, Penshurst, Victoria, private communication.
2. Queensland Irrigation and Water Supply Commission, *Report on Mareeba-Dimbulah Irrigation Project*, 1952.
3. W. Beckett, *Royal Central Asian Society Proceedings*, 1953.
4. Federazione delle Associazioni Scientifiche e Techniche, Milan, October 1965.
5. Merendi, *Banco di Roma Review*, November 1957.
6. Professor Isaac, University of Durham, private communication.
7. K. A. Ingersent, *Economic aspects of farm irrigation*, University of Nottingham, 1964.

Note: All brought to 1974 values using the Index of table 5.1.

a. Labour requirements, ½ to 1 man-hour per acre-inch (103 m³), representing 25–30 per cent of total cost. Capital charges, including depreciation and maintenance, computed at 11½ per cent.

b. Plan for irrigating 70 hectares (i.e., enough to keep one man fully occupied) using both dam and well water. Cost of equipment (i.e., excluding dam and wells) $400/hectare of which 100 electrical (power cost, 1·1 c/kWn); 20 per cent of total cost is for 500 hrs of managerial supervision, costed at £1/hr.

c. Capital costs were 117 c/m³—probably a world record.

d. Water supply is obtained from Qanats, or nearly horizontal adits to aquiferous rock in the hills. Currency is converted on the price of wheat in 1950: 400 tomans = 1 ton wheat = $67, un-dated to 1974 with the index. Capital cost of water supply is 1·51 c/m³ of capacity, and the owner's charges are at the rate of ½ per cent for maintenance plus 10 per cent net return. This cheap water is wastefully used.

e. An early study (International Institute of Agriculture, June 1913) gave a similar cost of Venetia, and for southern Italy (Campania) 5·3, of which 4·65 was the cost of an ox-lift of 12 m.

f. includes 0·5 cost of pumping for 30 m.

g. For a large pump of 638 m³/hr, costing £2,200, assumed 10 per cent capital charges and 2,000 hr/yr utilization (i.e., 0·06 c/m³). Power costs 1·33 c/m³, and labour 0·60 c/m³.

CHOICE OF FURROWS OR SPRINKLERS

Most irrigation schemes considered so far distribute large quantities of water through furrows. Where smaller quantities of water have to be distributed to valuable crops (for example, potatoes or tobacco) in a comparatively short time, spray distribution is desirable at considerably increased cost per m³. Capital costs are higher, despite savings on levelling, and fuel costs per m³ are about double.[66]

For the United States,[67] a detailed comparison of furrow and sprinkler costs is available (see table 5.15).

Table 5.15. Furrow and sprinkler costs (c_{74}/m^3).*

		Sprinkler	Furrow
Fixed costs:	Depreciation	0·44	0·16
	Interest	0·19	0·30
	Taxes and insurance	0·07	0·07
Variable costs:	Water charges	0·37	0·37
	Labour	0·61	0·58
	Power	0·31	—
	Maintenance	0·16	0·31
Total		2·15	1·79

*Estimates for a 24-ha farm receiving 71 cm in 7 applications every 2 weeks in summer

Source: Claude H. Pair, *op. cit.* note 61.

However, a careful study in New Zealand[68] shows much lower sprinkler costs at about 1·0 to 1·5 c_{74}/m^3 (0·895 $NZ equated to $\$_{70}$ after allowing for estimated difference in machinery costs). The estimates relate to grassland requiring, according to the weather, intermittent irrigation, and are ranged in ascending order of capital costs in table 5.16. Interest and depreciation combined are taken at 15 per cent of capital cost. Water inputs

Table 5.16. Annual costs of spray irrigation in New Zealand ($\$_{74}/ha$).

Method	Labour	Pumping	Maintenance	Interest and depreciation	Total
Tow line	6·8	22·0	6·8	17·2	52·9
Hand moving	13·8	11·6	4·8	29·1	59·4
Side roll	5·1	10·7	6·8	30·6	53·3
Self-propelled					
Big gun	1·7	27·1	11·3	40·6	80·8
Centre pivot	1·7	11·3	14·4	43·4	70·8
Side move— tow combination	3·1	7·9	11·0	45·4	67·4
Self-propelling straight lateral	1·7	8·2	15·5	46·5	71·9

vary from year to year. At 50 cm average, these represent costs of 1·1–1·4 c_{74}/m^3, consisting mostly of fixed costs.

It is claimed[69] that, in accordance with modern agricultural theory, spray-irrigated crops require much more work to be done on them. A 'tractor-labour unit' is defined as a 90-hp tractor plus 20 man-hours, which suffices to disc 50 hectares. Two units of cultivation, 2 units of discings, 1 unit of planting, 0·75 unit for liquid ammonia, plus 0·33 unit for other fertilizer application add up to 6·08 units for 50 hectares. Under furrow irrigation, the same area would need an additional 1·5 units for levelling, 6 for ploughing, and 2·5 for 3 harrowings—a total of 16 units.

In the case of furrow irrigation, nearly three-quarters of the capital cost is for levelling. Maintenance is estimated at 2 per cent on hand-moved sprinklers, rising to 6 per cent when they are mechanically moved. These are estimates for a 24-hectare farm receiving 71 cm in 7 applications (every 2 weeks in summer). However, Wright Rain (private communication) found in the Sena Sugar Estates in Mozambique that furrow irrigation called for higher capital investment than spray and used 25 per cent more water to attain the same result.

Wittig[70] estimates capital costs for furrow irrigation (including, in the area in question, considerable levelling) as high as 1486 $\$_{74}$/ha; for sprinkler equipment 988 $\$_{74}$/ha. This is close to a Californian estimate for an 80-hectare farm.[71] The life of the equipment might be anything from 25 to 50 years, but with a shorter life for the pump and motors. This project was planned for a well depth of 30 m; a depth greater than 45 m is probably uneconomic. Total cost, including capital charges, may range from 0·9 to 1·7 c_{74}/m^3.

In California, where much less levelling is required, the capital cost for furrow irrigation is estimated at only 472 $\$_{74}$/ha.

Another estimate for Israel[72] gives costs at the end of the delivery line of the National Water Carrier varying from 4·4 to 6·5 c_{74}/m^3, according to assumed rate of interest, to which must be added 1·7–3·0 for distribution (at the exchange rate prevailing before 1967).

Spray irrigation costs in Italy[73] were estimated at 3·1 c_{74}/m^3 on the plains but as much as 11·6 in hilly country. In Australia[74] costs were estimated in the 2·5–4·7 range (of which 25–30 per cent was labour cost).

Further evidence from Australia[75] gives 0·5 c_{74}/m^3 as costs of reticulation and maintenance, without any capital charges, for the very large dam on the Ord River in north-west Australia. The cost of pumping from permanent water at Alice Springs in central

Australia is given at 0·14 c_{74}/m^3, without any charge for the labour, which, it is estimated, would be otherwise unoccupied. At Carnarvon in Western Australia, on the other hand, pumping costs are estimated at 2·36 c_{74}/m^3, and the available supplies limited. Purchases of additional water rights from the graziers who are now using them would represent an interest charge of an additional 3·3 c_{74}/m^3. The cost of water from a proposed new dam at Rocky Pool would be 6·4 c_{74}/m^3.

Considerable further economies may be possible with drip or trickle irrigation, which gives little or no encouragement to weeds, does not interrupt farm operations, and requires only polythene pipes 1–1·75 cm in diameter as against the 12–15 cm diameter aluminium pipes required for spray irrigation. A permanent system can be laid in an orchard at a cost of 350 $\$_{74}$/ha as against 2170 for permanent sprinklers. Where water is scarce, the distribution losses of 50 per cent in furrows, 25 per cent with spray, and 10–15 for drip or trickle systems are extremely relevant.

More generally, it is to be expected that furrow irrigation will be preferred in a situation where there is relatively level land, abundant land and water supplies, and where labour costs little and is unskilled. Conversely, sprinklers and trickle systems will be preferred where land is uneven, slopes are steep, labour is scarce with high opportunity cost, where there is no scarcity of capital, and also where soils are extremely porous. As time passes, it is to be expected that irrigation will be extended to cover more irregular land and that water will become more scarce. In these circumstances, future extensions to irrigation will increasingly involve pipe distribution systems and application by sprinkler or trickle.

INSIGHTS FROM EMPIRICAL STUDIES

As with most items of capital expenditure, irrigation costs tend to increase over time with inflation. Of course, such cost increases are more apparent than real, because it is to be expected that the costs of competing investments increase at a similar rate. However, given the recent rapid technological advance in certain aspects of irrigation development, it is conceivable that the relative costs of irrigation are falling. Such an expectation is likely to be contradicted by the increasing development costs of marginal systems that have been described in this chapter. New projects must occupy less favourable sites and thus incur increasing unit costs. To some extent it is possible to alleviate these disadvantages by technological advances and economies of scale in important parts of the system (for example, storage); however, despite these factors, the over-all system costs are increasing.

Primitive pumping methods, using hand labour and animal power, are still practised in many parts of the world. Indeed, given the continued growth in population and the absolute limits on the amount of land and water and the increasing cost of industrial inputs and fossil energy, it is to be expected that primitive pumping may continue or increase in importance. This spectre of hundreds of thousands of smallholders lifting small amounts of water by means of hand pumps is already a reality in Bangladesh.

All the best diversion sites and, to a lesser extent, dam sites are occupied. Before new projects of this nature are undertaken, rehabilitation and modernization of the vast, existing irrigation networks are recommended. The costs of such programmes are about one-third of that of new projects, although rehabilitation costs may rise considerably if extensive field drainage and reclamation is necessary.

Partial budgeting (analysing part of the system using financial budgets) is a powerful and often-neglected technique for providing cost insights. This is demonstrated in this chapter with regard to the furrow vs sprinkler controversy. It is shown that, other things being equal, sprinklers are likely to be preferred when water is scarce, when capital is relatively plentiful, when land levelling is required, and when labour is expensive.

Cost information is always difficult to interpret because of the commonplace distortions to market prices. In the case of irrigation such difficulties are compounded because of the joint nature of costs in the multiple-purpose projects. This chapter has shown that throughout the world there is a very wide range of costs for irrigation development. Clearly, many projects have cost levels which preclude efficient use of resources. In various circumstances, irrigation development is pursued for non-economic reasons. This is legitimate, but the fact that irrigation is being used to generate employment, to facilitate regional development, or to enable self-sufficiency to be achieved in a particular commodity should be made explicit. And cost analysis is necessary to ensure that means of attaining the given goal or goals involving the least cost are being followed.

CHAPTER 6 ECONOMIC RETURNS TO IRRIGATION

CRITERIA FOR SUCCESS

Before 1950, irrigation schemes were judged largely on their financial merits. Criteria such as the financial productivity test applied in India were the main check on the merit of a project. This approach stemmed from a Select Committee of the House of Commons back in 1879, which stated that 'the financial results of works of irrigation are, in the opinion of the Committee, the best test of their utility'. The irrigation project had to meet, from revenue, simple interest on the original capital charge, plus annual working expenses. Pressures to recognize indirect benefits were largely resisted, though the rate of simple interest was occasionally lowered.

In the next two decades, the concept of economic efficiency was refined, and the use of social cost-benefit analysis became widespread. In this period, two criteria held sway: net contribution to national income and, because of the continuing power of Treasury officials, finance. However, although this approach indicated numerous socially productive projects, which for various reasons were financially unattractive, these twin criteria have proved to be too narrow. In the last decade, a variety of goals related to employment creation, income of low-income groups, regional development, and problems concerned with balance of payments have been used to supplement the goals of financial and economic efficiency. Often these objectives are conflicting; for instance, economic efficiency may conflict with regional development.

This complication of goals and criteria by which to judge results makes it much more difficult to assess the merits of individual proposals for irrigation projects or working schemes. Satisfactory economic and financial returns, though clearly desirable, are not necessary and certainly not sufficient in many circumstances. Nevertheless, this chapter is largely concerned with economic and

Notes and references for this chapter begin on p. 262.

financial issues. Even if multiple and conflicting criteria are to be applied, economic and financial matters have practical importance. For example, there is always a least-cost way of creating employment, and in a mixed economy this has to be financed.

GENERAL PROPOSITIONS

Although large numbers of feasibility studies for irrigation projects have been carried out in the last 20 years, few have been implemented. Almost all studies recommended that construction should be undertaken and that the scheme was financially and economically viable. The fact that few schemes have been implemented stems from two reasons. Firstly, irrigation requires resources such as foreign exchange and skilled engineering manpower, both of which have other, at least equally productive, uses. Secondly, the cost estimates have typically been underestimated, and the rate of uptake, yield levels and product prices have been over-estimated. In short, a typical feasibility study is generally found in the light of experience to be an over-optimistic document. Conversely, most new irrigation schemes have proved to be uneconomic—though they have other merits.

As previously mentioned, these general propositions are difficult to verify because of the paucity of *ex-post* evaluations. However, as will be shown at least for large-scale schemes (over 10,000 ha), what little evidence there is supports these statements.

International comparisons of returns present similar problems of valuation to that of comparing costs, which was discussed in the previous chapter. In the simplest analysis, the marginal returns to a cubic metre of water may be measured in the actual weight of additional crop or livestock product obtained. Net returns differ, however, from gross, in respect of additional expenditure on seed, fertilizers, cultivation, labour, and so on, for the (larger) irrigated crop. In low-income countries, these expenses are usually comparatively small.

RETURNS TO FARMERS

What is irrigation water worth to a farmer? It would be simpleminded to expect a single answer. We may expect widely different answers according to climates and according to crops grown. But we have not yet reached the end of the question.

Even when we are dealing with a single climate, and a single pattern of farming, the principles of economic analysis still tell us that we must look for a demand curve, and price elasticities of demand (defined, where q is the quantity demand and p is the price, as $\frac{dq/q}{dp/p}$ where d is the calculus sign meaning 'a small change'. Thus price elasticity is the proportionate effect on demand of a change in price). If we assume that a farmer is intelligent and well-informed, with his objectives uncomplicated by risk attitudes or social constraints, we can estimate the amount of water for which he (or she) can find remunerative use. This amount will depend upon the prices of his products, and of other inputs, and on the price at which water is offered to him. To take extreme cases, there must be some price so high as to compel him to do without water altogether; on the other hand, if it comes gratis, he will probably waste it. Within this range, the farmer has considerable freedom of choice as to the quantity of the kinds of crops which require supplementary water that he will cultivate; and also, for a given crop, the amount of water which it will receive.

In the simpler stages of economic analysis, economists have tended to assume unchanging price elasticities of demand for any one commodity. It is now clear that this is not generally the case, and that price elasticities themselves may change considerably.

The demand curve for water may be measured in the first place empirically, by observation of the amounts demanded by farmers under varying circumstances; but also in a more sophisticated manner by linear programming. In this case, the agricultural economist, armed with all the information about costs of labour, fertilizers, and so on, for each unit of crop production, each crop's responses to water, and the expected returns from it, sets out to tell the farmer precisely how much of each crop it is worthwhile for him to grow, with varying water prices and the expected total net return.

Where rainfall is uncertain, irrigation can increase the average returns, but from the farmers' viewpoint it may be the reduction in the risk of crop failure which is more important. Similarly, public decision makers may be interested to know the effect of various sequences of weather patterns on the rate of return. Carruthers and Donaldson[1] developed an investment model to simulate and assess the impact of possible weather sequences upon rates of return to sprinkler irrigation of tea in Bangladesh. With 500 experiments, they obtained the results set out in fig. 6.1. These show that even in the driest area (Chittagong), there is a 5

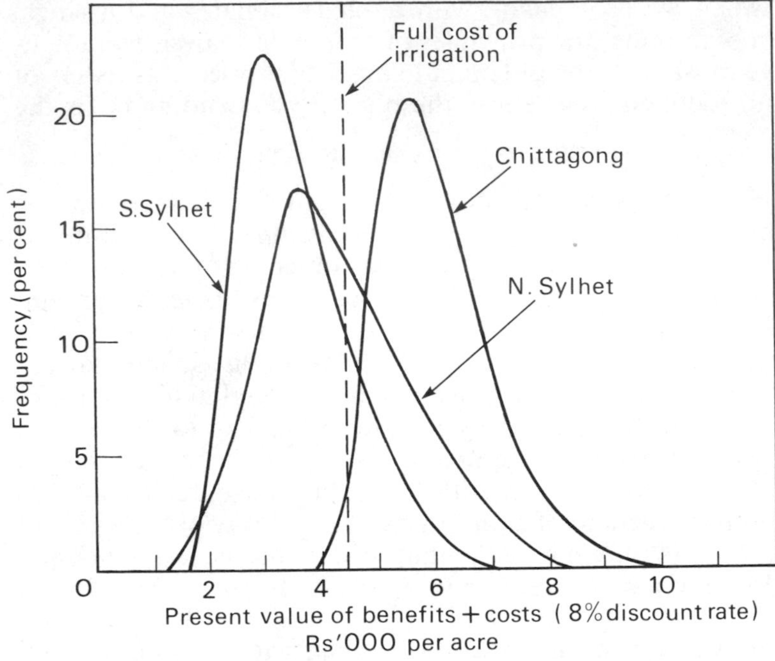

Figure 6.1. Benefits and costs of irrigating tea in Bangladesh.

per cent chance of a sequence of wet years after installing sprinklers, which will result in failure to reach an 8 per cent return on capital. This type of information, which enables the project analyst to retain the implication of a full range of technical assumptions to a late stage in decision making, is vastly superior to the normal single-value estimate of returns.

The most direct measure of the economic value of irrigation water is found under those (very rare) circumstances when farmers are able and willing to exchange it among themselves. Such a situation arose[2] in north Colorado (a dry but fairly cool area) in 1959, and it was found that the prices ranged from 0·35–0·7 c_{74}/m^3 early in the season, and 0·6–1·13 late in the season. Anderson attempted to make an independent estimate of the value of water by linear programming. His lowest figure was 1·28 c/m^3 for low-value crops grown on good soil. There seems to have been some 'buyer's rent' in the Colorado prices—that is, water came on to the market in times of unusual abundance, and was purchased by buyers who would in fact have been willing to pay a much higher price for, at any rate, a substantial part of their supply.

Valuing the rupee at $0·25, water pumped from tubewells in

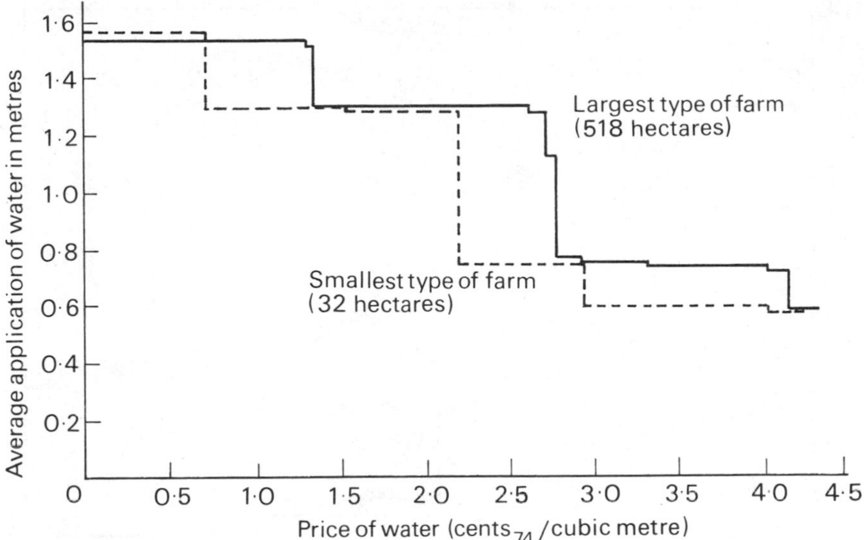

Figure 6.2. Demand for irrigation water, cash-crop farms in Tulare County, California, estimated by linear programming, 1963.

West Pakistan is sold at 1·8 c_{74}/m^3, the price falling to 1·13 as more wells are sunk and competition increases.[3] Valuing the Indian rupee also at $0·25, an enquiry at the village of Danda in Gujarat[4] in 1968 showed five wells at work in the village, each delivering about 100 m^3/hr. They had been installed fairly recently, and the water was still selling at 1·9 c_{74}/m^3.

A much more ambitious application[5] of linear programming to this problem was made in Tulare County, California, where cotton, beans, and sugar beet are the principal irrigated crops (see fig. 6.2.). For supposed farms of various sizes (and for the whole country by combining these estimates), programmes were worked out on the assumption of varying water prices, showing the precise points at which the use of increased quantities of water became worthwhile. No great differences were found between the largest and smallest farms. For the county as a whole, a price elasticity of 0·19 was found at prices up to 2·1 c_{74}/m^3, and of 0·70 at higher prices. (The figures include all water other than rainfall, whether pumped or drawn from surface flows.)

Linear programming was also undertaken for a 4-ha farm in West Pakistan by Gotsch.[6] Converting the rupee at 25 c, fig. 6.3 below shows average water input, measured over the whole area of the farm, at varying prices (the author also gives in detail the seasonal pattern of demand). It is interesting to see that, in spite

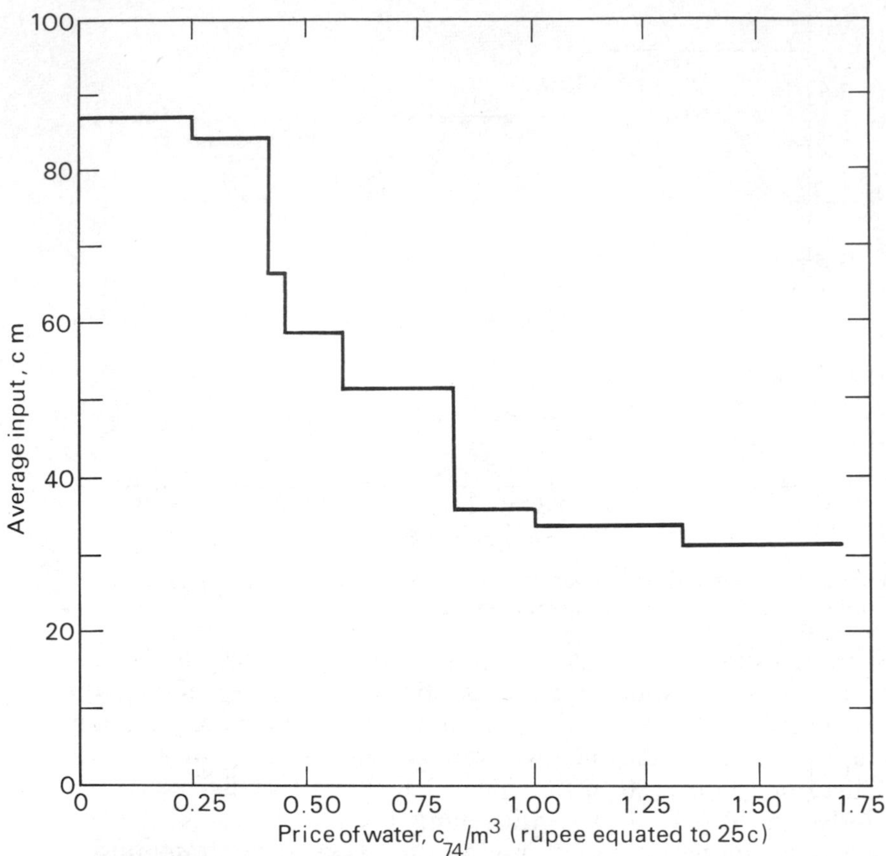

Figure 6.3. Demand for irrigation water on 4-ha farm in Pakistan.

of the small area of the farm, the amount of water demanded at a given price is much less than in California.

However, the provision of irrigation water, on a small farm of given area, does have highly significant effects in increasing both the demand for labour and its marginal productivity (see fig. 6.4).

Another programming study was undertaken[7] in the irrigation area near the mouth of the Rhône. This brought out the interesting result that the demand for irrigation water was greatly influenced by the labour supply. Land had been divided up into holdings each containing 15 ha capable of irrigation, with a view to growing fruits and vegetables. Where the labour supply was confined to the farmer and his wife, even if both were working full-time, they found it impossible to make a living out of vegetables and had to grow some cereals. In this case, with the price of water at $3{\cdot}0\,c_{74}/m^3$, they applied on the average only

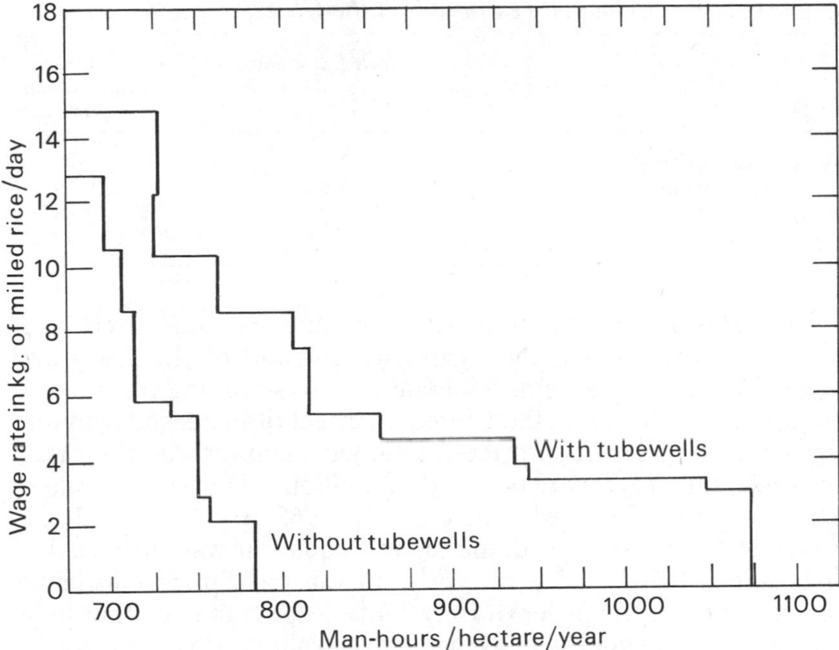

Figure 6.4. Labour requirements and wage rates on tubewell and non-tubewell farms in West Pakistan.

Source: op. cit. A programming approach to some agriculture policy problems in West Pakistan by Carl Gotsch.

11 cm, and earned a miserable income of 3000 francs/yr. If the price of water went appreciably higher, they would use considerably less, and their income would become negative. Alternative assumptions were then made about labour availability. It was found that the same results were obtained whether it was assumed that the farmer could have the services of only one permanent worker, or of one permanent worker together with five casual workers in the busy season; and whether or not the farmer required a minimum income of 15,000 francs for himself. The surprising result was obtained that consumption of water would remain virtually unchanged over a price range from 3·0 to 8·75 c_{74}/m^3.

An earlier study by Dawson[8] was designed to throw light on price elasticity of demand for water, and is shown as in table 6.1 below, for certain other areas. Water requirements vary, of course, with the extent and distribution of rainfall, and with evaporation. Climatic factors being given, Dawson estimated that price affected water use with an elasticity of about 0·5.

Table 6.1. Price elasticity estimates in the USA.

	Costs (excluding distribution) (c_{74}/m^3)	*Average application (m)*
Ainsworth, Nebraska	1·52	0·41
Texas High Plains	1·93	0·28
Antelope Valley, California	2·28	1·53
Eastern states	0·95	0·13

The Hawaii Agricultural Experimental Station[9] used programming to estimate the aggregate demand of the fruit- and vegetable-growing area at Molokai at prices varying from 0·9 to 4·2 c_{74}/m^3. In this case, the estimated effect of increased fruit and vegetable supplies upon prices in the local market was also taken into account. Aggregate demand, in million m^3/yr, was estimated at 3·3 minus 0·273 (price in c/m^3) in 1965. By 1970, with the estimated expansion of demand, the equation was proposed of 6·23 minus 0·302 (price in c/m^3). In this case price elasticities appear to be of the order of only 0·1 to 0·15. It may be that price elasticities are generally low with high-value crops, where payment for water represents a small proportion of gross sales.

Upton[10] programmed 100 ha in Botswana with 96 cm available water and 1350 mandays/month labour maximum (at 0·35 rands/day wages, but for 5 months of the year there was a labour shortage, in which the marginal productivity of an additional man would have been 4–12 times his wage!) The net return in this instance was 1·9 c_{74}/m^3.

Moore *et al.*[11] prepared a programme for the entire Imperial Valley area of California (190,000 ha irrigated) for four different degrees of salinity—so that, *inter alia*, they could estimate the trade-off between reduction of salinity and greater water input. Rate of plant growth was assumed to be a function of mean moisture stress in the root zone. The work called for a linear programme with 500 equations for 3 soil types, 3 irrigation treatments, 4 salinities, 3 leaching fractions, and 3 farm sizes. Lettuce was not to exceed 10 per cent of the area (otherwise the whole US market might be flooded), cotton not to exceed 9 per cent, sugar beet 12½ per cent (because of allotment laws), and alfalfa or fallow not to be less than 30 per cent (on conservation grounds). The main results are shown in table 6.2 below.

All these programming studies indicate what ought to be the water utilization pattern of rational farmers. In practice, few farmers achieve the potential returns. This is mainly because the

Table 6.2. Marginal return to water of varying salinities.

Average water input over whole area (excluding 5 cm rainfall)	*Marginal return (before paying land rent c_{74}/m^3) Total dissolved solids (parts per million)*			
	480	*960*	*1280*	*1920*
cm				
0–8	19·2	10·7	7·80	6·20
8–46	5·55	5·67	5·10	4·28
46–93	2·05	2·34	1·92	1·32
93–183	1·47	1·24	1·27	1·27
183–283	0·33	0·39	0·34	0·37
283–312	0	0·16	0·20	0·29

linear programming specification typically understates the real nature of the constraints to maximum returns or the influence of harmful exogenous factors, such as weather. On occasions, programming solutions under-estimate the returns actually achieved. However, typically, programming solutions exceed the realized returns.

RETURNS IN TERMS OF WHEAT EQUIVALENTS

Attempts can be made to standardize returns to facilitate comparative analysis by expressing all net returns in terms of world prices. Most countries have some degree of protection for internal prices of agricultural products. When comparing countries with various degrees of protection, it is convenient to work in terms of world prices. This procedure is not without difficulties because the world price is often difficult to ascertain. The world market is a residual market and therefore liable to wide fluctuations in price given relatively minor shifts in supply (for example, monsoon failure in India) or demand (large grain purchases by the Soviet Union). There are further complications that include fluctuations in exchange rates, lack of information on the quantity traded at a particular price, and unknown quality differences.

Standardization can be further attempted by expressing prices of all commodities in terms of wheat equivalents. The use of wheat equivalents is more valid in developing countries where grains represent a large proportion of agricultural output and where input costs are small in proportion to the value of gross output. In economically advanced countries, neither of these conditions occurs and net returns should generally be expressed in money terms.

In 1973–4, the average price for wheat imported into the United Kingdom was \$109·60 per metric ton[12]—in other words, 1 cent would purchase 0·09 kg wheat. In 1958–60, wheat prices were lower, at \$67 per metric ton at European ports, so that 1 cent should be expressed as 0·15 kg wheat. At that time sugar was equivalent to 1·27 times its weight in wheat (in 1973–4, 3·3 times), ginned cotton 8·65, butter 12·2. In California in 1960, sugar would have been sold at 50 per cent above world prices. However, we can reckon that Tulare County agriculture prices were only 15 per cent above world prices, hence 1 cent = 0·13 kg wheat. Approximate estimates made by Tulare County officials in 1967[13] of gross production value less current costs of cultivation and harvesting indicated a net return of 3·2 c/m^3 for cotton, 1·6 for barley or lucerne, 0·8 for grass. The price of irrigated land was about 1450\$/ha higher than that of adjacent unirrigated land used for grazing, implying a net revenue of 90 \$/ha/yr post-tax, or 165 \$/ha/yr pre-tax. Assuming an average water input of 1·2 m, this implies a net return of 1·3 c/m^3 or, using our index, 2·2 c_{74}/m^3.

An estimate of the marginal productivity of irrigation water, using production functions, was made by Hopper[14] for the village of Senapur in the Ganges Valley in India (where well water is abundant, if power is available to lift it). The marginal productivity of a cubic metre was found to be 0·25 kg wheat or 0·36 kg barley (the prices of the different crops in the area were found to be well adjusted to their marginal costs of production).

The lower of the two prices quoted as now being paid for water in Pakistan (see page 195) corresponded to no more than 0·070 kg/m^3 of wheat, 0·072 of maize, or 0·049 of rice (presumably milled), at prices prevailing there. One may perhaps expect that in the Pakistan Punjab, where both geographical conditions and the availability of machinery have made water easier to obtain, marginal productivity may already be much lower than in the Ganges Valley. This is seen to be the case (see table 6.3). Ghulam Mohammad[15] gives the average area of farms without tubewell irrigation as a little over 11 hectares in both districts which he studied (Multan-Montgomery, with low rainfall and water-table at 7·5 m depth, and Gujranwala-Sialkot, with higher rainfall and water-table at 3 m depth). Under the conditions of the Punjab, where irrigated land can be obtained very cheaply, farmers with tubewells operated more land, and double-cropped (to the extent of 50 per cent of their total area in the land with higher rainfall and 30 per cent in the drier land), until their water supply stood at about 1 m/yr per unit of area cropped. (They probably did not need all this, and sold some to their neigh-

Table 6.3. Marginal returns to water in West Pakistan (tons wheat equivalents per cropped ha).

	Higher rainfall area		Lower rainfall area	
	Before tubewell irrigation	After tubewell irrigation	Before tubewell irrigation	After tubewell irrigation
Areas operated (ha)	11·4	17·0	11·4	22·2
Area cropped (ha)	14·4	25·7	11·3	29·2
Tons wheat equivalent/cropped ha:				
Gross product	1·52	2·04	1·52	2·04
Debit fertilizer	0·04	0·08	0·04	0·09
Debit labour	0·44	0·27	0·38	0·25
Debit bullocks ('000 rupees)	0·80	0·51	0·82	0·46
Net product (debiting above inputs only)	0·24	1·18	0·28	1·24

bours.) Accurate information about increased yields on tubewell irrigated land is not available, and Ghulam Mohammad makes approximate estimates. He also estimates the manner in which the crop pattern is likely to be reorganized when tubewell water is available. Separate estimates are made for areas whose predominant crops are cotton and rice respectively, but the monetary results are not very different—gross returns are about 570 R/ha/yr without and 765 R/ha/yr with tubewell irrigation. (It must be remembered that the land without tubewell irrigation depends upon canal irrigation, but with inadequate water supply.) He finds fertilizer inputs raised from 51 to 114 kg/ha/yr of ammonium sulphate equivalents. This was available in Pakistan at a subsidized price of 130 R/ton at farm, but its true cost (judging from the experience of similarly situated countries) can be estimated at twice that. Ghulam Mohammad also gives the inputs of labour and bullock power; one of the great advantages of tubewell irrigation is that it makes possible the fuller use of these resources, which were formerly under-occupied. A man's labour is valued at 400 R/yr, and a pair of bullocks at 1600 R/yr. Finally, rupees are converted into wheat equivalents on the local price at that time of 375 R/ton.

If we assume that the use of tubewell water averaged a depth of 0·75 m on all cropped land—that is, 7500 m^3/ha, then returns measured in wheat equivalents are seen to be 0·125 and 0·128 kg/m^3 for the higher and lower rainfall areas respectively. In this area, circumstances are admittedly unusual, with additional (unirrigated) land to be had cheaply. The marginal product is much below Hopper's figure, but above the price paid for water ($0{\cdot}070 kg/m^3$).

However, there is more to it than that. Ghulam Mohammad writes:

> When a farmer saves or borrows 6–12,000 rupees and installs a tubewell his whole outlook on agriculture as a business changes. He wants to grow more valuable crops, to apply fertilizer, and to use other modern inputs to increase his income . . . Tubewells already installed are having a powerful influence on the saving habits of neighbouring cultivators, and most of them are planning to have their own tubewells. So far the only outlet which the farmers had for their savings was purchase of land or construction of houses. The price of (irrigated) land being extremely high, returns on investment were low and there was less incentive for saving. Now for the first time the farmers have got a low cost investment opportunity which yields extremely high returns.[16]

Falcon and Gotsch[17] found that the availability of tubewell water raised *net* income of 5-ha farm in Pakistan from 2097 to 2877 rupees (5·14 to 7·04 tons wheat equivalent).

NET INCREMENT FROM IRRIGATION

The estimates based on linear programmes or production functions, which in effect allow the cultivator to change his cropping pattern as his water supply changes, give results of greater interest than simple comparisons of the unirrigated and irrigated yield of a given crop. Most of our information is in this latter form; but, before we go on to examine it (table 6.5), we may consider a 'common-sense' programme for an Iraqi farm of 12 ha prepared by Jewett for Hunting Technical Services.[18]

For conversion to grain equivalents the ratio 20 dinars/ton wheat is used. Before the new scheme, the typical farm received (from rainfall and irrigation) about 25 mm/month April–September, and 60 mm/month in the winter, or about 0·5 m/yr; under the proposed irrigation scheme it will receive 0·96 m/yr, fairly evenly distributed. The returns are calculated net of seed and other materials which have to be purchased. Labour requirements, at present 249 man-days/yr, will rise to 535—no change is made for this, as it is assumed that partially occupied labour is available. If the additional labour were charged, the rate, from experience in

Table 6.4. Returns to irrigation in Iraq, 1961.

	Area (ha)		Crops (tons)		Net return (tons wheat equivalent)	
	Present	Proposed	Present	Proposed	Present	Proposed
Barley	4·2	2·0	4·4	2·8	1·76	1·12
Wheat	2·0	4·0	1·6	4·08	1·60	4·08
Berseem	–	3·0	(animal products)		0·82	3·17
Cotton (unginned)	0·25	1·5	0·2	1·44	0·60	4·32
Summer fodder	–	0·375	(draught animals)		–	–
Rice	0·125	0·75	0·16	1·2	0·24	1·80
Vegetables	–	0·375				1·50
Sesame	–	0·625		0·5		1·25
Total					5·00	17·24

other low-income countries, would probably be found to be a little over 3 kg wheat/man-day or, say, 1 ton in all, leaving a marginal net return of 0·94 ton/ha or 0·20 kg/m^3 of added water (1·02 tons/ha if no charge is made for labour). This budget was drawn up in the mid 1950s. Mitra (note 42, chapter 5) shows that nearly two decades later rehabilitated and reclaimed farm units of 7·5 ha produced more than 4 tons of wheat per hectare, more than four times what was anticipated. Furthermore, the labour requirement for an annual cultivated area of 11 hectares is only 374 man-days/yr because of the higher level of mechanization achieved. Net returns to labour and management have increased to 7 tons of wheat equivalent on unreclaimed farms (2 tons more than 1950s) but the reclaimed farms only yielded 14 tons of wheat equivalent which is more than 3 tons less than Jewitt expected.

Comparative gross and net products for a number of crops capable of being grown on newly irrigated land in southern Italy were calculated by Wittig[19] and are shown in table 6.5. The high returns from orange growing are obtained only with considerable skill and are also, of course, subject to interest on capital while the trees mature. The average rent paid, in wheat equivalents, was 625 kg/ha. To this must be added water charges, which were equivalent to 525 kg/ha for a 30-cm input, and 675 kg/ha for a 60-cm input (there was a substantial fixed charge and a small marginal charge). The combined level of rent and water charges—in effect, the marginal economic value of the water—measured in wheat equivalents, thus comes out at 0·38 kg/m^3 for an input of 30 cm, 0·22 kg/m^3 for an input of 60 cm.

A measurement of the marginal value of water precisely specified in time was made possible by the work of Anderson and

Table 6.5. Gross and net returns to irrigation, Sibari Plain, Italy. Kg/ha wheat equivalent (converted on the comparative Italian wheat price of 62·5 lire/kg).

	Wheat	*Maize*	*Lucerne hay*	*Sugar beet*	*Tomato*	*Orange*
Gross product	3000	3750	3750	5625	7500	20,000
Fertilizer	325	450	275	550	525	1050
Seed	250	225	75	250	175	5500
Insecticide	125	150	100	250	325	500
Labour and equipment	825	1425	1325	2400	4250	4000
Net product (before charging for water)	1475	1500	1975	2175	2225	8950
Water input (cm)	0	30	60	40	60	40
Returns in wheat equivalent kg/m^3						
Gross		1·25	0·62	1·41	1·25	5·00
Net		0·50	0·33	0·54	0·37	2·24
Value of 1 kg crop in terms of kg wheat	1	0·83	0·31	0·19	0·25	1·00

Maass[20] in Colorado, where fairly continuous irrigation is necessary during the summer. Approximate estimates from discussions with experienced farmers were made, of expected reductions in yield as a consequence of failure to apply water at certain critical periods. The calculations relate to the *gross* input of water—it is estimated that 50 per cent is lost before reaching the soil root zone. The values of gross product lost are assumed to be equally applicable to net product, that is, the reduction in crop is assumed not to lead to any savings in the costs of other inputs. The results are presented in table 6.6.

It must be borne in mind that the internal US price of wheat at this time was about 6·6 c/kg. Expressed in wheat equivalents, therefore, the order of magnitude of the values shown ranges from 0·1 to 3 kg/m^3. Some of the differences within the columns are interesting. Maize is exceptionally responsive to water at the end of June. Sugar beet, on the other hand, shows a fairly uniform demand throughout the season. Potato, a shallow-rooted crop, and also cotton show their highest response to water in the hottest season, but not so sharply peaked as maize.

The Green Revolution stemmed largely from improved cultivars of wheat and rice which responded to modern inputs (notably fertilizers), crop protection, and irrigation. Where such a package of inputs is applied jointly, a complementary response is achieved. This phenomenon of a more than additive response to improved inputs applied jointly has important implications for policy. It means that the highest marginal return at current levels of application is likely to be achieved in favourably endowed areas. Such a strategy runs counter to most current emphasis, which is upon reducing gross inequalities through concentrating resources upon the poorest regions, farms and farmers.

The phenomenon of complementary response was discussed in chapter 2. It would lead us to expect higher marginal returns to water in richer countries and better endowed regions where other constraints to production have been reduced or eliminated. This is, in fact, evident from table 6.7, presented in the final section of this chapter. If the gross and net marginal returns for higher value crops such as potatoes, tobacco, fruit and vegetables, wine, and even cotton and maize are examined, it can be seen that no Third World country obtains returns comparable with rich countries. In fact, no Third World country in table 6.7 obtains a gross return greater than 1·1 kg wheat equivalent/m^3, or about 12 c_{74}/m^3. Returns to irrigation water seem to follow the principle that 'unto everyone that hath shall be given'. High-value crops, grown by skilled, specialist producers, may show very high returns per

Table 6.6. Marginal values of irrigation calculated from loss of gross product as a consequence of water withheld (c_{74}/m^3).

Two-week period beginning:	Lucerne	Beans	Maize	Small grains	Cotton	Sorghum	Sugar beet	Potato	Apple
April 1			2·1	16·6					
April 15						12·2	14·3		
April 24	3·7	5·1	5·1	4·3		5·1			12·8
May 13				20·5		9·6	14·3		
May 27	3·3	5·2	3·7	2·4		3·3		17·8	
June 10				18·2		11·2	17·8		
June 24	8·9	12·2	1·9	23·1	4·2	11·2	23·8		
July 8	2·8	6·1	3·8		23·3	4·2	11·2	22·0	
August 5	5·4			14·0	3·3	11·2	17·8		
September 2				12·2			9·4	7·2	
September 16						9.1		7·2	
Gross values[a] \$/ha	320·2	379·7	38·5[e]	211·7[b]	1419·2[d]	332·5[d]	934·5	1318·0	1435·0[c]

All values up-dated to 1974 values using index (Table 5.1, page 119).

a. Expected on well-irrigated land at prices of 1954–8 (reference: *Colorado State University Agricultural Experiment Station Technical Bulletin*, no. 70, p. 26).

b. Barley.

c. Assumed 250 bushels/acre at Colorado price of 2·21 \$/bushel.

d. Californian values for 1950–9 to represent expected yield from irrigated land.

e. Mean of grain and silage.

cubic metre of water (though it must be remembered that such crops may only have a limited quantity of water for a short period in each year, and that overhead costs per cubic metre may therefore be high). But irrigation yields far less return in poorer countries.

This phenomena of greater returns in the more developed economies has its mirror-image with the construction costs which, despite much lower labour costs, tend to be higher in poor countries. Deepak Lal notes that in road construction bulldozer machine output was one-third to one-quarter of what manufacturers specified as obtainable.[21] This shows the importance of obtaining and using data on the actual field performance of machines in order to assess the economics of alternative construction methods.

In considering the high marginal net return shown by maize growing in the United States or Italy, it must be remembered that the circumstances are different from those in India. The maize-growing areas in the United States and Italy have an adequate winter rainfall and some summer rainfall; a certain amount of additional watering at the time when the grain is forming in the middle of the summer has an exceptional marginal effect in counter-acting the effects of drying winds; this applies also to wheat and cotton. Most maize growing in India is timed so that, in any case, this grain formation coincides with the period of heavy rainfall in the summer monsoon. It is the basic supplies of water still stored in the soil after the winter which the Indian maize grower lacks in comparison with the American or Italian. Replenishment of water supplies for maize growers by artificial means, therefore, should not be expected to yield so high a marginal return in India. Too frequent cultivation in India may also lead to some loss of stored water.

There are many advocates of irrigation who, even if the costs of a particular project exceed the expected economic return, will nevertheless urge the adoption of the scheme. They do not urge adoption because they seek to exploit public romanticism or because, as politicians, they like making speeches about large dams; they urge adoption because they believe that irrigation schemes bring indirect economic benefits, in the form of secondary employment, and additions to public revenue, which outweigh the direct loss.

Secondary economic benefit might accrue from an irrigation scheme designed for some particularly impoverished rural area in an under-developed country, where resources were under-utilized, labour under-employed, and the area capable of substantially increased activities without the importation of more resources of labour. However, as was pointed out in chapter 1, all investments in these circumstances will have secondary effects, and irrigation is unlikely to be vastly superior in generating secondary net benefits.

Camacho and Bottomley consider that secondary benefits (and costs) are worth estimating. For Guyana, using an input–output model, they show indirect financial benefits equivalent to the direct benefits and suggest an additional worker is employed outside the scheme's boundary for every worker employed within it.[22] It seems likely that this is the correct order of magnitude. Bell and Hazell find that on the Muda Irrigation Scheme in Malaysia, for every dollar of value added on the scheme, 75 cents of value added was generated in what they term 'downstream' effects.[23] However, excess capacity did not exist and to obtain this required additional investment in ancillary buildings, housing, vehicles, rice milling, and so forth. Despite the good quality of the data and their sophisticated methodology it is not clear what is the net return to secondary activity nor who gains what income from this activity. If future research reveals that there are important differences in the distribution pattern of secondary benefits toward poorer or, as is more likely, richer persons, this will be an important finding given the current concern with equity aspects of public investments.

Bergmann and Boussard devote a whole chapter of their book[24] to assessment of indirect effects but conclude that in present circumstances there are theoretical and practical difficulties in estimating them. They survey the theoretical literature and note a variety of interpretations of what is and is not an indirect effect. They point out the extreme difficulty of setting out certain economic relationships clearly. The practical problems arise because, even if an impeccable theoretical framework were available, the data and methods of calculation would usually be lacking.

So far as the developed countries are concerned, this argument may have had some validity in the 1930s when there were a great many unemployed, and transport and other public services were obviously under-utilized. Its use in the 1950s and 1960s revealed a grave intellectual obsolescence. At that time, the economist had to debit against what farmers, local traders, and others might produce on an irrigation settlement the value of what they might have been producing without irrigation, or producing elsewhere, and only credit the net difference; and he had to debit the full charges for use of transport and other services.[25] Today, in the early 1980s, with widespread structural unemployment, even in industrialized countries, we have a situation somewhat similar to the 1930s. Irrigation can take its place alongside other useful public works (for example, tree planting or environmental improvement) designed to generate social benefits from under-employed resources. However, it has no special merit, and therefore it should be subjected to the normal political, financial, and economic criteria by which public investments are assessed. There are few, if any, instances where irrigation development will more efficiently employ otherwise unused economic capacity than will alternative public works.

Table 6.7 facilitates a general consideration of available information on gross and net marginal returns to irrigation. In this table, all products are expressed as wheat equivalents. Net marginal return is estimated after deducting all additional costs incurred, including labour, which is assumed to be available (though this is by no means always the case). The table was prepared by Colin Clark for the second edition of this book. Although all the information (with the exception of the Bhakra discussion) refers to the period before 1969, it has been retained for two reasons. First, the data clearly indicate higher gross and marginal returns to irrigation in rich countries and this feature, which is discussed below, is still valid. To the extent that the higher returns now possible, following agronomic advances, are derived from the joint application of several inputs, including high levels of capital and skilled management, this divergence can be expected to continue or even grow. Second, the data are derived from a wealth of statistical sources, some of which are no longer accessible, and they have an historical interest. In a period of rapidly fluctuating exchange rates the wheat equivalent unit selected has a new appeal as a basis for international comparisons of agricultural efficiency.

The most generous responses are shown by potatoes, in Europe as well as in the United States. These are responses, however, to comparatively small quantities of water, using costly equipment which stands idle for most of the year, on heavily fertilized crops, in the hands of exceptionally skilled growers. Similar considerations apply, though less strikingly, in explaining the high yields from tobacco, cotton, and tomatoes in the United States, fruit, vegetables, wine, and flax in France, and fruit in Italy and Australia. In the case of maize, the already high-yielding crops in the United States give a much higher marginal response to water than the best French or Italian results. Among the cereals, the high returns to wheat in France are outstanding.

In the low-income countries, the marginal returns to water also appear to be low. A well-cultivated plant of good strain seems better able to take advantage of additional water than can a poorer plant. In West Africa the *Office de Niger*[25] was obtaining (in rice and cotton) only 1060 kg/ha of wheat equivalent in 1962. The average depth of watering was not stated. But the net return appears to be only about 0·1–0·2 kg/m³.

Table 6.7. Global review of returns to irrigation.

Country	*Crop or Product*	*Conversion factor (kg wheat/ kg crop)*	*Irrigation water input (cm)* Specified	*Irrigation water input (cm)* Not specified-approximately estimated	*Gross marginal return (kg wheat equivalent/ m³)*	*Net marginal return after meeting additional costs (kg wheat equivalent/m³)*
Australia	Butter fat	14·7	27		1·11	0·22
	Butter fat					0·94
	Grass (dry weight)	0·39			0·39	
	Dried fruit				0·75	
	Canned fruit				1·85	
	Rice				0·17	
	Sorghum					1·21
	Wheat					0·94
France	Flax	6·1	9·5		2·56	
	Fodder crops			40	0·62	0·47
	Fruit and vegetables			40	3·55	2·85
	Maize	0·75	25·3		0·55	
	Maize on sandy soil		24·2		1·55	
	Potato	0·65	7		11·54	
	Sugar-beet	0·15	23·3		1·14	
	Wheat	1		30	1·07	0·88
	Wine	0·85		30	2·56	1·45
India	General: Senapur					0·25
	Orissa					0·24
	Bihar					0·45
	Bombay					0·3–0·5
Iran	Wheat		220		0·045	
Iraq	Wheat			50	0·12	
Israel	Cotton (unginned)	2·4	50		0·41	
	Groundnut (in shell)	0·15	1·10		0·99	
	Sorghum	0·60	50		1·10	
	Sugar-beet	0·15	50		0·99	
Italy	General:					
	Campania, 1913		63		1·32	
	Venetia, 1962				0·09	
	Central Italy, 1962				0·96	
	S. Italy, 1962				0·66	
	Beans	1·76		50	0·60	
	Cattle and mixed	2·73		25	0·15	0·12
	Citrus (S. Italy)	0·48		50	2·03	1·17
	Grass					0·13

Country	Crop or Product	Conversion factor (kg wheat/kg crop)	Irrigation water input (cm) Specified	Not specified-approximately estimated	Gross marginal return (kg wheat equivalent/m^3)	Net marginal return after meeting additional costs (kg wheat equivalent/m^3)
Italy *cont.*	Hay	0·24		50	0·24	
	Lucerne (dry matter)	0·47			0·70	
	Maize (Piolanti)	0·75			0·82	
	Maize (Grüner)	0·75		50	0·30	
	Maize (Cofani)				0·58	0·31
	Maize second crop				1·10	0·54
	Peaches				1·86	0·53
	Potatoes				5·20	0·82
	Sugar-beet	0·068				0·26
	Sugar-beet (sucrose content)	0·92			0·49	
	Wheat			50	0·26	
Jordan	Wheat			50	0·47	
	Tomato	0·37		50	0·23	
Lebanon	General			50	0·67	0·38
New Zealand	Grass (dry weight)	0·385	64		0·27	
Nigeria				50		0·08
Pakistan	Cotton (unginned)	2·41	50		0·24	
	Cotton		88		0·03	0·03
	Cotton		85			0·07
	Millet		47			0·06
	Rice unmilled (E. Pakistan)	0·80	34		0·45	
	Rice (milled)	1·2	120		0·12	
	Sugar-cane	0·13	100		0·13	
	Sorghum	0·6	20		0·08	
South Africa	Sugar-canc (sucrosc content)	0·92				
	Tongaat	0·92	52		1·10	
			66		1·40	
	Illovo	0·92	71		1·05	
			105		0·90	
	Do. with abudant K_2O	0·92	71		1·29	
			105		1·08	
	Tongaat First Ratoon	0·92	36		1·76	
			46		1·40	
	Second Ratoon	0·92	75		1·32	
			84		1·48	
United Kingdom	Barley (25 kg/ha N)	0·65		10	0·24	
	Barley (50 kg/ha N)			10	0·08	
	Early potatoes	2	6		11·3	
	Sugar-beet (as sugar)	1·03	10		1·78	
USA	Butter fat (Tennessee)	14·9	53		1·06	

Table 6.7. Global review of returns to irrigation (continued).

Country	Crop or Product	Conversion factor (kg wheat/ kg crop)	Irrigation water input (cm) Specified	Not specified-approximately estimated	Gross marginal return (kg wheat equivalent/ m^3)	Net marginal return after meeting additional costs (kg wheat equivalent/m^3)
USA cont.	Cattle liveweight:					
	Illinois	12·1	31		0·46	
	Georgia		20		0·54	
	Cotton ginned (Georgia)	6·5	14		2·31	
	Hay: Massachusetts	0·47	15		0·68	
	Nebraska	0·40		30	0·57	
	Maize: Georgia	0·75	18		1·68	
	Missouri		14		1·43	
	Virginia		18		2·43	
	Milk: Virginia	1·3	28		0·53	
	Potatoes:					
	Long Island	0·65	7		5·83	
	Wisconsin		8		9·50	6·55
	Sorghum: Georgia	0·60	8		1·03	
	Soya beans: Missouri	1·3	12		1·02	
	Sugar-cane (sugar content): Hawaii	0·92	66		0·06	
		0·92	99		0·03	
	Sweet potato: Georgia	0·3	25		1·03	
	Tobacco: Virginia	8·0	18		5·90	
	Tomatoes: Georgia	0·37	15		3·57	

Australia

Butterfat.[1] Results for Murwillumbah in the Northern Rivers district of New South Wales, where the rainfall is usually good. A time series analysis of three farms in this area over ten years suggested results about twice as high, but Waring pointed out that these advantageous marginal results were only to be obtained in the intermittent dry years. Excessive moisture, in his opinion, reduced dairy output. He also made an important point in cow psychology: 'Stock allowed limited access to palatable feed at regular times will often stand waiting for long periods beforehand. There is a disinclination to forage as diligently on unattractive feed when something better is expected to be offered.' He gave the expected butterfat output (1 kg butterfat makes 1·2 kg commercial butter) from sown pastures at 90 kg/ha/yr, with the possibility of raising it by 250 kg by 45 cm irrigation. Another authority[2] claims that sprinkler irrigation can raise this yield as high as 336 kg/ha/yr of butterfat.

Allowance is made for additional output of calves and culls (10 per cent of gross product). Of the difference between gross and net return 62 per cent is accounted for by labour—on the assumption that the farmer is not able to provide any more himself.

Third entry. It is interesting to see that a result similar to the above can be obtained[3] by a regression analysis on variations in the rainfall in the month of April, usually the driest month of the year and a critical period for the growth of grass. The estimated return, however, is double on fertilized land.

In estimating economic returns it must be remembered that butter in Australia sells at about 25 per cent above world price (the coefficient of 14·7 quoted in the table is at Australian price)[4]. A more recent study by Bird and Mason[5] in the same area of the detailed accounts of fifteen farms over some years does not give the water input, which varies greatly from year to year, but assuming 43 cm average, the median gross return was found to be, in wheat equivalents, 0·41 kg/m^3 (or an increment of butter fat of 145 kg/ha, similar to the Milk Board's estimate). Additional costs, however, were incurred (reverting to wheat equivalent units) of 0·23 kg/m^3, plus 0·13 kg/m^3 cost of irrigation which leaves little net return from the operation, if butterfat is reckoned at world rather than Australian price.[6]

Grass.[7] Results from Gippsland, an area in Victoria with comparatively cool climate.

Fruit and Rice.[8] Canned fruit has very considerable expenses of production to be charged against the gross return.

Sorghum and wheat yields are estimated[9] to be directly proportional to the water content in the top 90 cm of the soil. This measure includes stored water at the beginning of the season and excludes rainfall which runs off. The maximum possible storage, even in heavy soils, is estimated at 15 cm.

1. Waring, *op. cit.*

2. G. Mason, 'Towards a productive function for supplementary irrigation on North Coast dairy farms', *Review of Marketing and Agricultural Economics,* **31,** June 1963.

3. New South Wales, *Milk Board Journal,* February 1959.

4. Waring, private communication.

5. J. G. Bird and G. Mason, 'The economics of spray irrigation on the far North Coast of New South Wales', *Review of Marketing and Agricultural Economics,* December 1964.

6. ——, ——, 'Budgets for plant irrigation on a typical Richmond Valley dairy farm', *Review of Marketing and Agricultural Economics,* March 1965.

7. Bank of New South Wales, *Spray Irrigation,* 1956.

8. Murrumbidgee Irrigation Authority publications.

9. Bell, *Journal of the Australian Institute of Agricultural Science,* December 1957.

France

Fodder crops, fruit and vegetables, wheat, wine (Lower Rhône Valley).[10]

Flax, maize, potato, sugar-beet.[11] Supplementary irrigation June–September in dry years only.

Unirrigated yields tons/ha:

Wheat	0·8	Maize	5·9
Wine	3·0	Maize, sandy soil	4·3
Flax	1·35	Potato	34·0
Fodder crops	1·2	Sugar-beet	41·0
Fruit and vegetables	5·0		

10. J. Klatzman, *La Localisation des Cultures en France,* Institut National de la Statistique et des Etudes Economiques, 1955.

11. M. Hallaire and P. Tabard, 'Supplementary irrigation under climatic conditions of the Paris region', *Academie d'Agriculture de France* (Compt. Rend. 47617), 29 September 1961.

India

In view of the importance of the subject to India, there has been regrettably little work done on the economics of irrigation.

Indian Statistical Institute Planning Unit,[12] p. 62, gives marginal returns from irrigation for each Indian state expressed in terms of food-grain equivalents. The quantities of water used are unfortunately not stated. The table is attributed to 'Directorate of Economics and Statistics', but nothing is said about the methods by which the results were obtained.

A Gokhale Institute publication[13] in 1948 reviewed eight villages in which the proportion of canal land irrigated varied from nothing to 36 per cent (there was also some well-irrigated land, but limitations of ox-power to draw the water result in this having much less effect on productivity). These results can be—somewhat hazardously—extrapolated to estimate the gross and net marginal productivity from complete irrigation. The data are the averages for 1938–9 and 1939–40 (95 rupees = 1 ton wheat). Gross product data as given were re-defined to exclude manure, seeds and fodder produced and used on the same farm, but before debiting purchases of these commodities.

In another Gokhale Institute publication, Sovani and Rath[14] estimated for Orissa, a region with a natural rainfall of 1·5 m and where, therefore, on the face of it there should be little need for irrigation, that irrigation could raise the proportion of land double-cropped from 22 to 67 per cent.

Professor Gadgil of this Institute in 1958 headed a group who prepared *Evaluation of the Benefits of Irrigation Projects* for Planning Commission Research Programmes Committee, examining five areas in different parts of India. The definition here of net marginal product is slightly defective in that some small payments for hired land and capital (but not imputed rent and interest on own property) have been debited (434 rupees = 1 ton wheat).

Gadgil's results were as follows (net marginal product representing the addition to factor income in the agricultural sector after paying for all additional inputs from outside the agricultural sector, and agricultural produce used up in production; further columns then show this figure after debited (i) hired labour, (ii) hired and family labour):

	Tons wheat equivalent/ha		*Do. after debiting*		*Net marginal product indicated from difference in land prices*
	Gross marginal product	*Net marginal product*	*Hired labour*	*All labour*	
Indian Statistical Institute					
Lowest State (Mysore)	0·42				
Median	0·62				
Highest State (Andhra Pradesh)	1·02				
Sovani and Rath, Orissa	1·63	1·18			
Gadgil, 1958					
Tribeni Canal, Bihar		2·26	2·16	2·19	0·28
Cauvery—Mettur Scheme, Madras		0·65	0·46	0·48	0·26
Damodar Valley, W. Bengal		0·16	0·05	0·02	
Gang Canal, Rajasthan		0·13	0·06	0·06	0·21
Sarda Canal, Uttar Pradesh		0·03	0·02	0·05	

Source: Professor Gadgil, *Evaluation of the Benefits of Irrigation Projects,* Planning Commission Research Programmes Committee, 1958.

	Percentage of area canal-irrigated	Tons wheat equivalent/ha Gross product	Net product	Do. less labour
Gokhale Institute 1938–40				
Four villages	0	0·40	0·31	0·07
Belpimpalgaon	7	0·35	0·28	0·05
Ozar	21	0·87	0·62	0·06
Yesgaon	31	2·34	1·54	0·58
Rahate	37	2·37	1·39	0·52

Source: As above

In no case do we have information on the average quantity of water supplied. If we assume this to be ½ m, then each hectare receives 5000 m³, and Hopper's figure of 0·25 kg/m³ would imply 1·25 tons wheat equivalent/ha net marginal product. This is well above the Indian Statistical Institute estimates, of the order of magnitude of the Sovani-Rath estimate, and lies between the first and the second of Gadgil's districts in his 1958 study. The very low figures for the Sarda Canal area are explained in terms of its being easy in that area to obtain well water for land not covered by canal irrigation; the low figures for the Gang Canal and Damodar Valley are harder to explain. When the column 'After debiting all labour' is higher than the previous this means that family labour input is actually reduced in the irrigation areas, and replaced by hired labour. The calculations, based on rises in land values* (lower values were shown by the Punjab Board of Economic Inquiry)[15] assume that they represent 20 years' purchase of net incomes; high rates of interest are often found in India, but these do not appear to apply in the case of land, where security is regarded as fully adequate. The low net incomes indicated may be explained by the fact that by no means all the land in the area is fully covered by canal irrigation every year.

	*Expected net income from land, rupees/ha 1950–1 to 1956–7	1957–8
Old irrigated land	50	84
Bhakra Dam irrigated villages	40	43
Unirrigated land	15	23

It appears that there is some uncertainty as to whether Bhakra Dam will in fact deliver all the water expected. The irrigated–unirrigated difference of 61 rupees in 1957–8 represents 0·125 ton wheat/ha, and the earlier figures less than 0·1. By 1971 the uncertainty on water availability was confirmed. The dam can only be filled on a live storage capacity of 6787 × 10⁶m³ as some capacity has to be reserved for flood control. This storage plus the inflow during the depletion period (September 21 to May 31), which is 3455 × 10⁶m³, give aggregate availability in the depletion period of 10,242 × 10⁶m³. The irrigation demand is estimated at 11,500 × 10⁶m³ giving an aggregate shortfall of nearly 1358 × 10⁶m³ per year.[16]

However, in 1964 Harbans Singh published details of an ingenious scheme to incorporate tubewells into the system to augment scarce supplies at periods of peak irrigation demand. At this time irrigation demand exceeds hydro-electricity demand therefore spare electric power can be used to power tubewells placed in batteries alongside the canals. Releases from the reservoir can then be reduced without reducing farm supplies; the saved water can be used later when either irrigation or power is required. Because this procedure maintains a higher head in the reservoir, there is more power generated per unit of water.[17]

The Gokhale Institute studies of 1938–40 show a remarkable variance within a limited area (they were confined to what was then known as Bombay Province). They show that irrigation villages which are willing and able to make large inputs of fodder, fertilizers, etc. (see differences between gross and net products) can earn high net incomes. If we extrapolate to the situation of a village where all the land is canal-irrigated, we find net products of 3–5 tons/ha. In this region water inputs may be of the order of 1 m, implying 0·3–0·5 kg/m^3, still considerably higher than Hopper's figure.

12. Indian Statistical Institute Planning Unit, *Price Policy for Irrigation Undertakings: A Preliminary Study,* August 1961.

13. Gokhale Institute publication no. 17, *Economic Effects of Irrigation,* 1948.

14. Sovani and Rath, Gokhale Institute publication, *Economics of a Multiple Purpose River Dam.*

15. Board of Economic Inquiry, Punjab, Publication no. 78.

16. B. S. Minhas, K. S. Parikh, T. N. Srinivasan, S. A. Marglin, and T. E. Weisskopf, *Scheduling the Operation of the Bhakra system,* (Statistical Publishing Society, Calcutta, 1972).

17. Harbans Singh, 'Firming up of hydro-power by tubewells and apportionment of costs', *Irrigation and Power,* **21,** 3, 1964.

Iran

Beckett.[18] There appears to be a wastefully high input of 2·2 m of water. The land grows nothing without watering.

18. W. Beckett, Royal Central Asian Society Proceedings, 1953.

Iraq

Yudelman,[19] (dry yield 0·6 ton/ha).

19. M. Yudelman, 'Some issues in the agricultural development of Iraq', *Journal of Farm Economics,* **40,** 1, February 1958.

Israel

Experimental results.[20] Inputs represent entire water supply apart from pre-sowing irrigation. Response was approximately linear within this range of input, but further inputs beyond 50 cm gave no significant response.

Cotton and sugar-beet, see reference *21*.

20. D. Yaron, *Economic criteria for water resource development* (Hebrew University of Jerusalem, 1966).

21. Y. Lowe, 'Irrigation and settlement schemes', International Conference of Agricultural Economists, 1967.

Italy

Italian agricultural prices at that time assumed to be 25 per cent above world level.[22] The irrigation season lasted for 6 months, and yield was raised from 1·05 to 9·3 tons wheat equivalent/ha/yr. Some 30–40 per cent of the gross product was payable in rent.

See Vicinelli[23] for 1962 general results. Dry yields for the three areas were 1·39, 3·85, and 1·24 tons wheat equivalent/ha respectively. The cheapness of water in Venetia apparently causes it to be used comparatively wastefully—indeed, the yields for the three areas show a curious proportionality to the cost of the water which is 0·31, 3·0, and 2·15 c/m^3 respectively.

Grüner,[24] for beans, hay, wheat, and one datum for maize, pre-irrigation yields 0·8, 3·5, 0·9, and 1·5 tons/ha respectively. Other data from Tofani.[25] (The cattle and mixed farm holding of 7·5 hectares had a pre-irrigation output of 1·47 tons/ha gross, 1·24 net. The conversion factor here refers to cattle liveweight.)

Citrus data from Vicinelli.[26] Non-irrigated yield about 3 tons/ha. Additional costs include 15 per cent gross return on irrigation capital.

Data for lucerne hay supported by Merendi.[27]

Lucerne, last entry for maize, and sugar-beet, from Piolanti,[28] who conducted 12, 5, and 15 experiments respectively. The figures given here represent the yields at optimum water inputs. Yields are given in physical quantities. Interesting insights are provided by a comprehensive study by Antonietii, d'Alanno, and Vanzetti (Istituto Nazionale di Economia Agraria). Tables are presented for each province showing the usual input of water for each irrigated crop, and the expected increases in yields (measured in money) and in labour requirements. Labour was costed at the average earnings prevailing in 1961 (357,000 lire/man-year in the north, 250,000 in the centre, 184,000 in the south and 215,000 in the island: cost per man-day, when labour inputs were quoted in this form, taken at 1/200 of a man-year). Other deducible costs were taken at 17·5 per cent of the gross harvest (25 per cent for orchards). All values were converted to wheat equivalents at 68,100 lire/ton.

For general arable cultivation, the median of 11 provinces showed 50 cm of water input and a net return of 0·28 kg/m^3. Rice appears to be over-watered, with Lombardy and Piedmont using 3–4 m, with a net return of only about 0·1 kg/m^3 wheat equivalent. Sugar-beet, in three provinces, averages 0·35 kg/m^3 net return, lucerne, in three provinces, 0·33 kg/m^3 (average watering 80 cm), grass, in five provinces, shows an average net return of 0·14 kg/m^3. Excessive watering appears to be frequent here also. Tomatoes show 0·90 kg/m^3 (average watering 46 cm). Vineyards in six provinces, with a median water of only 15 cm, show a high median return of over 3 kg/m^3. In Lazio, spray irrigation of 7 cm yielded 4·9 kg/m^3, furrow irrigation of 23 cm yielded 1·5 kg/m^3. The median return from orchards in seven provinces was 0·61 kg/m^3 from 20 cm watering.

22. *International Institute of Agriculture Monthly Review,* June 1913.
23. 'Irrigation in Italy', *International Journal of Agrarian Affairs,* **3,** January 1963.
24. Grüner, *Bewasserungs Anlagen,* (Zurich, 1944).
25. *Genio rurale,* (August 1955).
26. Vicinelli, *op. cit.*
27. Merendi, *Banco di Roma Review,* November 1957.
28. Academia Nazionale di Agricoltura, September 1968.

Jordan

Davies.[29] Dry yields wheat 0·17, tomatoes 2·97 tons/ha.

29. H. R. J. Davies, 'Irrigation in Jordan', *Economic Geography,* **34,** 264–71, 1958.

Lebanon

Ward.[30] Dry yield £L 207 (i.e., 1·22 tons wheat equivalent)/ha. Ratio of net to gross from UN Relief Organization, Beirut,[31] from data for Syria, where gross output is similar to that in Lebanon.

30. G. H. Ward, 'Economics of irrigation water in the Litani Basin of Lebanon', *International Journal of Agrarian Affairs,* **2,** 5, June 1959.

31. U.N. Relief Organization, Beirut, *Quarterly Bulletin,* March 1958.

New Zealand

Woudt.[32] Perennial ryegrass and clover pastures, dry yield 10·7 tons/ha/yr.

32. B. D. Woudt, 'Recent advances in the irrigation of pastures in New Zealand', *World Crops,* **10,** 6, June 1958.

On irrigated light land farms in the Canterbury area, Stuart and Haslam[33] found, with average water input 200 cm/yr, net income after debiting for management, depreciation and 6 per cent interest on capital, of £7·3/ha as compared with £7·8 on unirrigated. 'Irrigation on these farms becomes merely a drought insurance rather than an income-earning investment' they wrote.

33. Stuart and Haslam, *Lincoln College Publication,* no. 6.

Pakistan

East Pakistan rice from Ghulam Mohammad.[34] East Pakistan is a high rainfall area, and inputs of water beyond 34 cm showed practically no further response.

Cotton (second entry) from Rafiq and others.[35] No additional costs of cultivation were incurred. The response on fertilized land was double.

Third entry for cotton, and millet, from Carruthers.[36] Eighteen per cent of the water input required for the cotton, and 23 per cent of the requirements for the millet, represented 'scare early summer water'. Conversion was effected at local prices (rupee = 2·19 kg wheat).

34. 'Private tubewell development and cropping patterns in West Pakistan', *Pakistan Development Review,* **5,** 1–33, 1965.

35. C. M. Rafiq, M. Alim Mian, and R. Brinkman, 'Economics of water use on different classes of saline and alkali land in the semi-arid plains of West Pakistan'. *Pakistan Development Review,* **8,** 23–34, 1968.

36. I. D. Carruthers, *Irrigation development planning,* (Wye College, University of London, 1968).

South Africa

Cleasby.[37] Experiments by Tongaat Sugar Company, Natal (in a high rainfall area). For plant cane water supply ranged from 1·9 to 3·1 m and for ratoon (second growth) from 1·0 to 1·9. A kilo of sugar is worth 1·27 kg wheat at world prices; the sucrose content of the cane 0·92 kg (Australian data used for milling costs).

Some interesting cross-relationships with fertilizer inputs are in progress. So far, it appears that the responses to nitrogen of ratoons are unaffected by additional water inputs.

37. Cleasby, *Africa and Irrigation,* (Wright Rain Ltd., 1961).

United Kingdom

Nix.[38] The results for early potatoes were obtainable only in dry years. For sugar, a distribution of results was given:

Increment tons sugar/ha	Zero or negative	Under 1·25	1·25–2·5	2·5–5	7·5	Over 7·5
Number of cases	6	9	3	4	2	1

38. J. S. Nix, Cambridge Farm Economics Branch Report no. 55; and *Agriculture,* May 1960.

USA

Department of Agriculture Year Book.[39] Also *Agricultural Engineering*[40] for grazing, and *The Agricultural Situation*[41] for Nebraska lucerne.

Hawaii data refer to additional water supplies above a high opening input of 381 cm.[42]

39. *US Department of Agriculture Year Book, 1955.*
40. *Agricultural Engineering,* March 1955.
41. *The Agricultural Situation,* February 1953.
42. Waterhouse and Clements, *Hawaiian Plant Record,* 271, 1955.

CHAPTER 7 CHARGES FOR WATER

THE FUNCTIONS OF WATER-RATE

Charges made for irrigation water are generally well below costs. There are many and varied reasons for this. The basic reason is that there are at least three functions for a charging system. These functions are *economic*, *financial*, and *social*.

Economic function

Because the economic function is to ensure that resources are efficiently used, the appropriate criterion is whether the price charged is equal to society's valuation of the resources utilized in producing the service. If the consumer price exceeds the marginal cost of providing water, full economic utilization is likely to be prevented.

In the case of old-established irrigation supplies, it is common for prices to be associated with historical costs. These are generally lower than long-run marginal costs because easier and cheaper sources were available and because of the effects of inflation. They may also be lower than short-run marginal costs because of political difficulties in increasing rates necessitated by either inflation or rising marginal costs. In this situation, water is cheaper to the farmer than the real economic cost. This may, in turn, lead to demand exceeding supply at the going price, to water shortages and demands for new, even higher-cost augmentation.

Financial function

The financial function is that the water-rates should cover the costs of the service. These costs include capital costs, operation and maintenance costs, and revenue collection costs. In this case inflation also has to be taken into account.[1]

Notes and references for this chapter begin on p. 264.

Social function

The social function is a mixed bag of policies and actions whereby water pricing may be used to promote income redistribution, economic stability, or to develop backward areas and encourage investment by beneficiaries.

INFLUENCE OF COST STRUCTURE

Irrigation service has many of the characteristics of a public utility. Studies of the pricing problems of public utilities have concentrated upon the general text-book case and on the costs of supply rather than demand.[2] It is clearly not possible to recommend irrigation water-rates without some simultaneous insight into the likely pattern of demand and the costs for varying levels of supply.

It was shown in chapter 5 that irrigation has an unusual cost structure. There are high fixed costs for such investments as source works and distribution facilities, and for any particular scheme there is a wide range of capacity where there are increasing returns to size. This is clearly demonstrable with pipeline and dam capacities and is also true for canals; for example, for pipelines laid costs are roughly proportional to pipe diameter, but carrying capacity is roughly given by the relationship $d^{0£2}$ where d is pipe diameter. However, there is a dynamic aspect to the problem. The decreasing cost/capacity relationship has to be balanced by the temporal pattern of demand, which is lowest at the beginning of the life of the project, and by the costs involved in having capital invested in excess capacity.

In addition to high fixed investment costs, there is a large part of recurrent cost which is not variable with amount of water consumed. For example, the costs of canal operator and maintenance staff are independent of the level of use. These are sometimes called 'indivisible costs'. For a gravity irrigation scheme, true short-run marginal costs are practically zero.

The classical economic argument for pricing is that price should equal marginal cost. This would signal to consumers the resource cost of incremental supply. Marginal cost pricing is considered to be economically efficient. However, if long-run marginal costs are falling, the operating agency will run at a loss. In the more normal situation, where long-run marginal costs are rising, pricing using marginal cost principles will generate financial surpluses. Experience suggests that such a policy is impractical for three main reasons. First, where long-run marginal costs are rising, the political will to use irrigation for revenue generation is demonstrably absent. Second, in developing countries, the system of market prices is so difficult to establish and, even so, in such a distorted economy, optimizing pricing with respect to these costs is no guarantee that from an overall national viewpoint marginal cost pricing is efficient. Third, the logic of marginal cost pricing suggests that, given different costs, different rates should be adopted for different projects. Furthermore, within a project, those farmers located at a distance from the source would have to bear higher costs than those near to the source. Where consumption over the year is uneven, farmers using water at the period of peak demand should, in principle, bear the full costs of supplying the peak. Such refinement has rarely been achieved in the much more easily regulated field of drinking water. In irrigation, the principle of 'postage stamp pricing' or an even charge across a country is generally adopted in spite of variation in marginal costs.

To summarize: first, for any given scheme over a considerable operational range, unit costs of irrigation water fall by increasing the size of the scheme; and second, unit costs per m^3 of water fall with an increasing level of utilization. Third, over-time, incremental scheme costs tend to rise. These three factors have considerable impact upon irrigation water-rate policy.

SUBSIDIES BY DESIGN AND DEFAULT

Conflicts among economic, financial, and social interests within government policy are seldom resolved in favour of finance. It has been shown that most of the low-cost irrigable sites have been exploited, and therefore high returns are a necessary precondition for financial success. Returns in practice have often been disappointing. Government may choose to subsidize irrigation up to the extent of a free supply, as in Thailand. Where irrigation is financed through general taxation, the expectation is that the benefits will be diffused throughout the economy and that the costs will be recouped through general taxation. Given the level of cost and uncertain benefits, this appears to be an extremely optimistic approach. Unlike other users of public infrastructure (such as transport systems) irrigation farmers are easily identifiable, direct beneficiaries of a production input. Because in most developing countries public revenue is one of the most valuable resources, it is judged that some form of water rate policy is advisable.

Usually, however, financial criteria are relaxed by default, the government being reluctant, for political reasons, to increase rates in line with inflation.

The rate of uptake of irrigation is generally much slower than is forecast. Sometimes public agencies, supported by economists, encourage full, early use of the irrigation facilities by promotional prices. However, it is by no means clear that there is an elastic demand for irrigation water. Lowering prices will have little or no impact if the effective constraint upon the use of irrigation water is a plentiful supply of rainwater, the absence of field channels or level land, scarcity of labour or farm power, or the ignorance of farmers of the techniques of irrigation agriculture. Where some measure of price elasticity exists, low initial water rates may encourage more rapid uptake.

Promotional pricing policies cannot be revised later to financially realistic targets unless the operating agency has considerable political support. However, given the political will, the relatively inelastic demand for irrigation water (once a system is established) will be an asset in raising revenue.

There are at least five possible economic reasons that might be used to support subsidized irrigation rates. Firstly, where increasing returns to scale in construction (decreasing average cost) have encouraged installation of a high level of installed capacity, which for one or more reasons is under-utilized, rates at, or close to, short run marginal costs are optimal. Secondly, and more generally, it can be argued that the beneficiaries of irrigation include not just the farmers but also, indirectly, many others. For example, there are firms supplying farm inputs and firms processing and marketing the outputs who will benefit from increased turnover. Similarly, it is possible that the incidence of benefits will lie in part with consumers of the output. It is possible to argue that the incidence of costs for irrigation should lie not just with the users but with all the beneficiaries.

Thirdly, the reason for financial subsidies is that often there are major distortions in the market prices for factors of production and outputs because of government interference, e.g. import duties, export taxes, or minimum wage legislation. The over-all effect of these distortions is to make the financial costs higher than their real (economic) costs. Various financial subsidies can be justified to help correct market distortions and to provide the appropriate economic signals to producers. A subsidized water tariff may be desirable in these circumstances.

Fourthly, very low irrigation rates may be found in countries with many old-established schemes. Costs of these schemes were initially low and, because they are now fully depreciated, only operation and maintenance costs have to be covered by rates. In these circumstances, it is usually not politically practicable to set very different rates for new, high-cost projects. Water rates within a country tend to be equal no matter what the cost of supplying water to the individual scheme (postage stamp pricing principle).

Fifthly, betterment levies, which are discussed later, are sometimes placed on the increase in capital value of irrigable land. In this case it would be double taxation to levy annual irrigation rates on a full cost basis.

BASIS FOR CHARGES

There are three basic approaches to setting charges. Rates can be related to the costs of providing irrigation, to the benefits to be derived from irrigation, or to some value judgement on the beneficiaries' ability to pay rates.

If a cost approach is used, rates can be set to recover capital and recurrent costs, or some proportion of them. In a low-income economy, the main problem with this is that such a focus may ignore both the level of benefits actually achieved (which may be low) and the poverty of the beneficiaries (which may be too abject to allow the desirable cost recoupment). Furthermore, when implemented, the impact of a cost recovery policy upon the level of consumption has to be taken into account.

An original and valuable approach to the problem of charging using a cost basis has been made by Wynn,[3] working in the Sudan. Wynn points out that marginal costs of water nearly always differ from average. He then quotes another American study,[4] which indicates that at least half of the present 1000 m m^3/yr used in irrigating the south coastal area of California would be uneconomic if users had to pay the marginal cost of incremental supplies now being drawn from the Colorado River. He therefore proposes that the fixed costs of the dam and canal system should be isolated, and recouped in the form of a levy on all land capable of receiving water from it—at the time when the land becomes capable of irrigation, not during construction (interest during construction is another element in fixed costs). Potential users at the far end of a long and costly canal would have a higher levy to pay; if the original designers of the system knew that such a rule would be put into operation, and that there were limits to what farmers could pay, we might be saved from some uneconomically long canals, as in India. Other costs should then be recouped as a charge per unit of water used. Where a simple barrage is capable of yielding a substantial water supply (as in the Sudan), Wynn suggests that only the estimated cost of such a barrage be included in the fixed costs to be recovered by the levy on land, and that the more costly water from the top of the dam should be charged at marginal cost. The marginal charges also would take account of the additional seepages and evaporation in long canals. The tendencies of governments to use water charges for purposes either of subsidization or taxation are economically undesirable, as in the case of the Sudan, which taxes cotton by making it bear the whole cost of irrigation, and subsidizes other crops by allowing them water free. Simpson, in a subsequent issue of the same journal, makes the interesting point that high marginal costs for maintenance are more likely to arise in the minor canals than in the major; the latter run at 'non-silting velocities'.

The limits upon a cost approach to water-rates is set not by the lack of an adequate theoretical framework but by the fact of extremely low farm incomes in many schemes and an almost universal lack of political determination to recover costs.

If the benefit approach is used, it is advisable to rate a *proportion* of the incremental benefits from irrigation (benefits with irrigation less benefits without irrigation) and not simply the net benefits from irrigated agriculture. In assessing incremental benefits, some allowance will have to be made for increased family labour consequent upon irrigation and other non-monetary costs. In an area with relatively good rainfall, the incremental benefits from irrigation will be quite small and high charges may deter its use.

If the criterion is ability to pay, then data on net farm incomes are required, stratified in order to indicate the incomes of the smaller farmers, those with unfavourable soils, or other low-income farmers. The basic procedure for determining the financial impact of benefit pricing is indicated in fig. 7.1.

First, the decision-making authority has to be persuaded to define 'minimum acceptable income'. It has then to define the basis for rating policy in terms of what proportion of benefits attributable to irrigation is taxable. Then an agricultural economist can forecast, on the basis of farm budgets or linear programming models, whether the scheme is likely to be subsidized and the extent of subsidy, or whether it will break even, or in rare instances, become a revenue-raising investment.

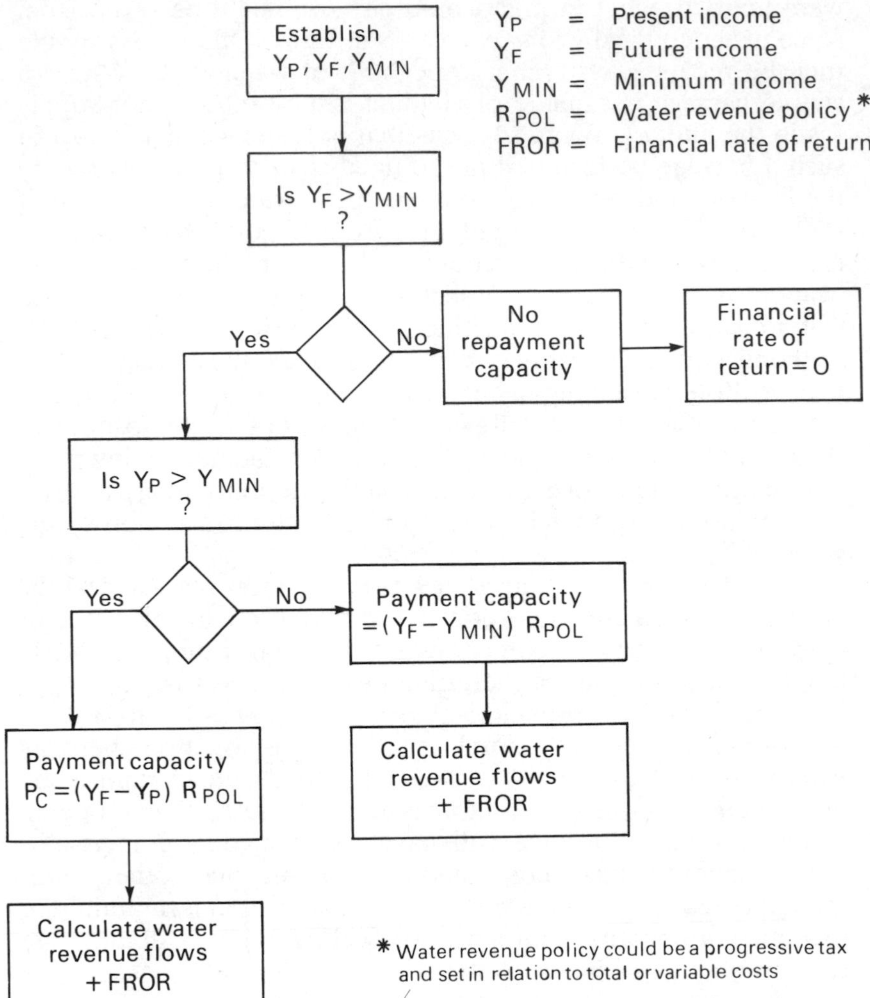

Figure 7.1. Procedure for assessing payment capacity using benefit approach

In selecting the basis for charges, public authorities will have to balance their financial needs, their policy towards financial independence of operating agencies, their approach to income redistribution issues, the extent of other subsidies or indirect taxes, and the practicality of enforcing the selected charging scheme.

In the final analysis, the setting of rates is a political decision. However, it is the responsibility of economists to ensure that politicians are aware of the economic, financial, and social consequences of their alternative choices.

CHARGING LEVIES

Once the basis of charge is decided, some measurable unit or charging vehicle has to be selected. This can be the amount of water used, the cropped area, the area and type of crop, or simply the income of the farmer. Volumetric measurement is technically difficult to ascertain for the relatively large discharge of irrigation works. An exception is where water is pumped from a constant head, because fuel costs can be used to derive a measure of water quantity supplies. Measures of discharge over calibrated concrete V notch weirs depend upon regular, accurate readings, and are not generally practicable for large-scale projects unless there are a few large farmers.

Volumetric supply has been used extensively and successfully in the United States and Australia. In Israel and Taiwan, volumetric supplies are made to co-operatives, who then allocate among smallholdings. In India, large sugar-cane farms in Maharashtra are charged on the basis of volumetric supply. In northern India and Pakistan, the constant discharge module (a slot exit gate for water from the canal which allows approximately the same discharge no matter what is the height of water in the canal) was introduced to facilitate volumetric charging, but there are numerous operational problems including module damage, module by-passing, extremely irregular canal flows, and disputes among the many small farmers sharing a particular module. These led to the abandonment of module charging in some areas and the introduction of an irrigation supplement to land revenue or an irrigation cess or rate based upon the area cropped. Where estimates of water used are obtained indirectly, it is absolutely essential that the water supply is reliable.

When the charging vehicle is the amount of water used, various types of tariff can be adopted. A two-part tariff with a fixed rate independent of quantity used and a variable charge for quantity has some financial advantage to the operating agency in that some revenue is assured whatever the influence of weather or other factors affecting demand. Sometimes the fixed charge is set to recover capital and the variable charge to recover recurrent costs. Although this may appear to be a neat solution, there is no economic rationale to support it. The variable charge may be constant, increasing, or decreasing with quantity. Similarly, where no fixed charge exists, the tariff can have varying levels for increasing quantities. Examples can be found where bulk rates are cheaper and where bulk supplies are more expensive. The appropriate form will depend upon the supply and demand conditions and the weight placed upon the various economic, financial, and social functions of water-rates.

Where area of land is used as a charging vehicle, there is the obvious advantage of simplicity and relative ease of measurement. However, there is no incentive to the farmer to use water judiciously. Indeed, in areas where water is scarce and where a land surplus exists, it may discourage development of the optimum area under irrigation.

Sagardoy[5] reports that in Israel a form of regional pricing is carried out in that regions are divided into low-cost regions (1·7–2·5 c_{74}/m^3) and high-cost regions where water is imported (3·7–15·0 c_{74}/m^3). Where low-cost supplies are found, farmers pay actual costs plus an additional tax which is equivalent to half the national average production cost (4·5 c_{74}/m^3). Where costs exceed the low-cost level, prices are in the range 2·2–3·4 c_{74}/m^3 and the rest is subsidized (up to 10 cents/m^3). Israel is considering using very high prices as a rationing device for the 20 per cent marginal supplies because of the dire scarcity of water within the economy.

Betterment levies

One device used widely in India and Pakistan to help finance projects is the 'betterment levy'. This is a tax on part of the increase in capital value of land that occurs following provision of an assured irrigation supply. It may be a single tax or spread over a number of years, sometimes as many as 7.

A betterment levy and high annual rates would, of course, be double-taxing the benefits. However, a reasonable procedure might be to have betterment levies related to capital costs and annual rates related to operation and maintenance costs.

For some uncertain reason there is often expected to be less reaction by beneficiaries to a betterment levy than to high annual water-rates. Perhaps landowners are thought to be in close touch with land values and therefore to realize that a betterment levy, if fixed at a realistic level, still allows part of the increased land value to accrue to the owner. In the Ghulam Mohammad Barrage Command in Pakistan, the betterment levy was 50 per cent of the increase in capital value of land that occurred with the introduction of controlled irrigation, payable over 5 years.

Financing the betterment charges from farming can be a problem, particularly as charges fall in the early years of a project when agriculture is relatively unproductive and there are many competing demands for finance—for example, field channels, field levelling, and drainage.

Resistance to betterment levies can occur even where there are low annual charges. For example, the charges[6] made for water from the Bhakra Dam in India have been purely nominal (0·02–0·05 c/m^3, in comparison with the higher, but still low, charge of 0·10 on old canals). (In this chapter charges have been adjusted by the inflation index, although most experience suggests that in practice few, if any, adjustments have been made.) Although the canal system of the dam was extended to grossly uneconomic lengths for political reasons, the attempt to collect a betterment levy 'after one year of collection of instalments had to be held in abeyance on account of wide-spread political agitation against it'.

In the land benefited by the Hirakud Dam, it was proposed to collect, by instalments, a betterment of 185–370 R/ha—only a small part of the capital cost. In Mysore (where irrigation costs are high) it was at one time customary not to begin an irrigation project unless the owners of two-thirds of the land benefited would agree to pay a betterment charge of over 370 R/ha. In modern India, however, it is considered that such methods are 'incompatible with democracy'.

If there is strong government enforcement of the betterment levy, there may be other advantages besides the obvious financial ones. The need for finance may encourage land sales. Land use is typically less intensive on large holdings. Given a ceiling upon ownership of land, if a more vigorous market in land can be developed, this will provide a shift of irrigated land ownership to smaller farmers, and more productive holdings will occur.

LEVELS OF CHARGES

Charges are seldom set high enough to recover costs. In a survey of 17 projects financed by the World Bank, anticipated recoveries averaged only 30 per cent of total costs.[7]

The charges made for canal water in India are purely nominal, as can be seen in table 7.1. In almost every case charges are less than 5 per cent of gross output; in many cases they are less than 1 per cent. Furthermore, given the apparent reluctance of state governments to tax agriculture effectively, and the widespread attempts to hold public sector prices to help fight inflation, it can confidently be predicted that water rates, as a percentage of the value of gross output, will fall rather than rise over time.

This situation is not universally approved in India. Indeed, the National Council of Applied Economic Research has recommended that water rates should be related to the incremental net

Table 7.1. Canal irrigation charges for major crops in India.

State	*Crop*	*Average gross output Rs/ha**	*Water-rate Rs/ha**	*Rate as gross output (%)*
Andhra Pradesh	Rice	1406	37·5	2·7
Bihar	Rice	553	37·5	6·8
	Wheat	818	22·5	2·8
Gujerat	Rice	1377	45–62·5	3·3–4·5
	Wheat	1417	45–62·5	3·2–4·4
Haryana	Rice	773	24·4	3·2
	Wheat	1262	14·5	1·1
Kerala	Rice	1556	12·5–25	0·8–1·6
Madhya Pradesh	Rice	1090	25	2·3
	Wheat	791	10–25	1·2–3·0
Maharashtra	Wheat	773	22·5	2·9
Mysore	Rice	1092	40	3·7
Orissa	Rice	NA	2·5–20	
Punjab	Rice	716	24·4	3·4
	Wheat	1353	14·6	1·1
Rajasthan	Wheat	1055	13·1–25	1·2–2·4
Tamil Nadu	Rice	1120	40–50	3·6–4·5
Uttar Pradesh	Rice	781	10–35	1·2–4·5
	Wheat	1041	10–30	1·0–3·0
West Bengal	Rice	1699	13·4–31·3	0·8–1·8

Source: Report of the National Irrigation Commission, 1972.
*Assume R1 = $0·25

income, that is, the difference between net benefits before and after irrigation. (It should be noted that this is not the usual test of net benefits, which is the difference between net benefits with and without irrigation.) They recommend a rate which may vary between 25 and 50 per cent of the increment.[8] This is equivalent to 4–15 per cent of the value of gross output.

The Irrigation Commission (1972)[9] recommends a crop basis for charges at 5–12 per cent of gross income. However, it also indicates other, possibly conflicting criteria—for example, rates should be within paying capacity of the irrigator; between regions with similar class of supply, rates should be equal, and the level of revenue from irrigation should be such that, in any state, no burden is imposed by irrigation on the general revenues. However, it is quite likely that 5–12 per cent of gross income would be beyond the paying capacity of many small farmers, and, even if collected, it would be insufficient to meet the irrigation costs.

Very low charges, between 0·05–0·10 c_{74}/m^3, are also found in

Pakistan. Ghulam Mohammad[10] records for the area, which he studies in the Punjab, higher charges averaging 0·17 c_{74}/m^3, slanted in favour of rice (0·09) and against the successful cash crop of oil seeds (0·47).

These represent charges for canal water. Rafiq[11] quotes a range of charges as follows (rupee equated to 25 c):

Canal water	0·14 c_{74}/m^3
Government tubewell	0·41 c_{74}/m^3
Private electric tubewell	0·42–0·84 c_{74}/m^3
Private diesel tubewell	up to 1·26 c_{74}/m^3

On the other hand, the Mwea-Tebere project in Kenya[12] charges K£25/ha/yr (halved in the first year). The irrigation authority will also undertake, amongst other things, land preparation and provision of seed, but makes a charge for them. Valuing the Kenya pound at $4, and assuming gross input to an average depth of 1 metre, this represents 1·7 c_{74}/m^3. This covered the full costs. In another project at Perkerra, the charge is the same, but is inclusive of ploughing, seed, insecticide, and spraying. In this project revenue covered only 48 per cent of expenditure. By 1976 water-rates had not risen although the cost and returns had increased substantially. Revenue from water-rates and cultivation charges failed to cover scheme salaries and wages.[13] At Perkerra they were less than 50 per cent of salaries and wages.

An Indian official estimate[14] shows an average depth of irrigation for wheat, peas, and fodder crops of 10 cm, of sugarcane 15 cm, and of rice 20 cm, with an average charge of 0·018 R_{74}/m^3 for the first four crops, and half that for rice. A rupee at that time would purchase about 2·2 kg of wheat. The charges clearly are far below the marginal productivity, of the order of magnitude of 0·25 kg/m^3; and supplies of water had to be rationed.

Irrigated land near Verona[15] first irrigated in the 1890s, used to be charged a rate equivalent to 30 per cent of the previous gross product; this charge forced many owners to sell their land. An improved set of canals operating entirely by gravity was completed in 1940. The land is watered for five months, beginning 15 April, at an average rate of 4 mm/day, or a depth of 60 cm in all (to this must be added an average rainfall of 1·3 mm/day during this period). The charge is 0·82 c_{74}/m^3, of which two-thirds represents operating and one-third capital charges. The charges made in the 1930s, before the completion of the gravity canals when the water required pumping, were twice as high in real terms. It may be added that much of this water is lost in a deep gravel sub-soil, and this, in its turn, creates a drainage problem in the Po Valley, a considerable distance away.

Water charges in central and southern Italy are still made per hectare and, assuming a consumption of 8000 m^3/ha, are equivalent to 0·6–1·2 c_{74}/m^3.

In Australia[16] charges are more than nominal, though still below cost, averaging 0·36 c_{74}/m^3 (20 shillings/acre-foot) in New South Wales and 0·24 in most of Victoria (though rising to 0·87 in Ṇyah and 1·05 in Tresco, where pumping costs were high).

In southern Italy,[17] subsidies meet 1·32 out of a total cost of 3·82 c_{74}/m^3.

According to Ringulet,[18] the Bas Rhône et Languedoc scheme in France, where the average input is 46 cm, charges an average of 3·77 c_{74}/m^3. This is an average based on 6·16 c_{74}/m^3 for the first 15 cm, 2·87 for the next 15 cm, and 1·23 for any further water. Sagardoy[19] finds a wide range of prices in France from 0·09 to 1·19 c_{74}/m^3 for flood systems, and 1·78 to 3·15 c_{74}/m^3 for modern sprinkler systems.

In Lebanon,[20] on the other hand (assuming average depth of watering of 0·75 m), farmers have been paying the full cost of 10·80 c_{74}/m^3 for water pumped from the Litani River. New schemes are proposed which will supply them, at a loss, at 3·5 c_{74}/m^3 for pumped water and 2·4 $c_{74}{\cdot}m^3$ for gravity-supplied water.

In Iran,[21] as already stated, water can be obtained from adits to aquiferous rock, and is usually supplied in return for a share of the crop worth about 0·27 c_{74}/m^3, which gives the owner a 10 per cent net return on his investment. In vegetable gardens in Tehran, price (calculated on the assumption of a water use of 6 mm/day, or 2·2 m/yr) rises to 3·2 c_{74}/m^3.

The productivity of Indian agriculture is not generally high. Nevertheless, the rates levied in some states are well below the returns to water. For sugar-cane growing[22] in Maharashtra, a range is made of 0·54 c_{74}/m^3; it seems clear that the marginal productivity of water is much higher than this, and that the water could have been better used on other crops. Even these low charges have been reduced.[23] In Gujarat, charges for state tubewell water were reduced from 0·59 c_{74}/m^3 to 0·42 for those who pay a deposit in advance (double for those who do not).

In the United States, irrigation charges vary widely. At 1974 rates, the variation is clearly seen in table 7.2 below. Renshaw[24] found that other users of water paid several times this level. The average and maximum values, adjusted to 1974 values, are shown in table 7.3.

Table 7.2. Charges for irrigation water in the United States.

Region	*Charge* (c_{74}/m^3)
Great Basin	0·113
North West	0·125
Upper Missouri	0·146
Upper Arkansas	0·158
Upper Rio Grande	0·226
Colorado River	0·295
Central Pacific	0·339
Western Gulf	0·860
Southern Pacific	1·270
Upper Quartile in Southern Pacific	2·400

Source: E. F. Renshaw, 'Value of an acre foot of water', *Journal of the American Waterworks Association,* **50,** 303–8, March 1958.

Price elasticity of demand for domestic water, apart from garden sprinklers, is estimated at 0·23.[25] Sprinkling, which constitutes 40 per cent of the demand in metered areas in the western United States (60 per cent in unmetered areas) is estimated to have a price elasticity of 0·7 in the western United States and 1·6 in the east. The corresponding income elasticities are 0·35, 0·4, and 1·5. In the American west, potential evapotranspiration on lawns is estimated at 57 cm/yr, and water actually applied to them at 35 cm. For the east, the corresponding figures are 38 and 18 cm. Urban water demands are relatively price-inelastic. Urban consumers have considerably higher value-in-use. This explains why cities are sometimes able to outbid agricultural users. McClellan[26] states that the city of Colorado Springs offered 63 c_{74}/m^3 for agricultural water. This is perhaps an exceptional case, for it is nearly 250 times the average charge for irrigation water in the United States, and many times the agricultural value-in-use. Koenig, in the same volume (page 321), quotes a

Table 7.3. Range of charges for water according to use, United States, 1974 values

	Charges in c/m^3	
	Average	*Maximum*
Domestic	15·20	35·8
Industrial	6·01	25·0
Irrigation	0·25	4·1
Power	0·10	0·9
Waste disposal	0·10	0·4

Source: As table 7.2.

case of a Texas town, in an emergency, paying more than twice this amount.

Among commodities in different markets, water must hold a unique position, due to the wide variation in its price. The practical importance of the wide divergences in prices for the same product with different types of consumers is clearly brought out in the following quotation.

> Industry does not have a fixed requirement for water but rather a variable demand which depends importantly on the price of water. Thus, depending on price, steel mills may demand from 6 m^3 to 290 m^3 water to produce a ton of finished steel; and power plants may use 0·006 to 0·77 m^3 to produce a kilowatt-hour of electricity.

In arid regions, irrigation agriculture pays typically very low prices for water and uses huge quantities. In California as a whole, 90 per cent of the water is used for irrigation. The Imperial Valley irrigator, for example, pays 0·16 c/m^3 for water, whereas Los Angeles and other cities in a nearby region are paying 2 c/m^3 wholesale of the Metropolitan Water District. As for the urban user, distribution costs raise the price to about 6·5 c/m^3, and these cities face very much higher costs for future increments of supply. Inefficiency and waste are strongly indicated when a wide divergence of prices for the same product exists—a difference much greater than the cost of transfer.[27]

Price is seldom used to ration scarce supplies. An exception to this generalization is found in Alicante, Spain, where water is auctioned weekly by the irrigation community and sold daily to other farmers by individual holders. Nevertheless Maass and Anderson[28] contend that in most areas of Spain and United States they studied, farmers refused to treat irrigation water like a normal economic good such as fertilizer, and various social or community pressures preclude water prices being regulated by laws of a free market. Water is generally sold at less than the price farmers would be willing to pay.

There can be very few instances where irrigation charges are said to be too high. Taylor, examining the Pekalen Sampean Irrigation Project in East Java,[29] found farmers gained annually about $\$_{74}$ 200 per farm from irrigation, almost wholly as a result of increased cropping intensity. Indeed net returns per hectare were actually lower on three of the four main crops when irrigated and non-irrigated crops were compared. Before a rehabilitation project was instigated farmers paid $\$_{74}$ 9·10 per hectare to local irrigation operators, of which \$2·00 represented main canal expenditure although a higher charge of $\$_{74}$ 6·75 was considered optimum. Following rehabilitation main canal charges were to increase from $\$_{74}$ 2·00 to $\$_{74}$ 10·85 per hectare, which he regards as a transfer of income from agriculture to non-agriculture. This is hard to sustain in view of the non-charging for irrigation capital investment which undoubtedly increases private land values. He justifies this stance using the indirect benefit argument. He claims the benefits are shared by government, businessmen, labourers and consumers. However, it is not clear whether they gain more or less than if alternative agricultural investments were undertaken.

Finally it is worth noting that it is not only in mixed economies that water rates present apparently intractable problems. James Nickum, an American 'China-Watcher', who specializes in water sector matters, reports evidence from articles published in China that water rates are not always paid in that country. (He also reports other familiar problems of irrigation, including that there is often a lack of ancillary facilities to irrigation such as drains, that unresolved maintenance problems exist, that there are design defects such as tubewell outlets being placed too high, and that head/tail conflicts on canals exist at times of peak water demand.)[30]

CONCLUSIONS

Water-rates offer a powerful instrument of policy which is usually neglected by public irrigation authorities. Rates are seldom equivalent to full costs; often they are below operating costs. In inflationary conditions, rates are seldom increased in line with inflation, often because of the misguided view that inflation can be stemmed by holding prices in the public sector. Users of irrigation water are easily identifiable and direct beneficiaries of public resources. Where the farmers are landowners, there is an immediate benefit in that the increased potential of earning income is reflected in an increase in the capital value of their land. Thus they may realize a private gain from public investment by selling land.

Unless farmers are below socially acceptable standards of living, it is judged that water-rates should be levied. Providing that a minimum subsistence income is assured and that there is some return to the increased effort and skills required of irrigation, farmers' rates should be set to indicate to the farmers that irrigation water is a valuable resource which requires judicious use. Short-run marginal costs (including collection costs) should be the minimum charge. Long-run marginal cost pricing, which in most conditions will result in a financial profit, has obvious economic and financial advantage, but because of social and political obstacles it is likely to be unworkable.

CHAPTER 8 PLANNING IRRIGATION DEVELOPMENT

Two related problems complicate the process of irrigation planning. The first is created by the large number of interest groups, each with very different affinities and approaches to the subject. These groups promote irrigation because it is consistent with their view of the needs and priorities of development. The second problem is that any development project has to satisfy a large number of objectives, some of which are consistent but others which are conflicting. Assessing the trade-off between various levels of achievement of multiple objectives is one of the activities in irrigation planning. Furthermore, in planning irrigation projects, a sequence of activities is observable. These are identified and assessed in the last two sections of this chapter.

INTEREST GROUPS

The main viewpoints which need to be considered are shown in figure 8.1 and include those of the agriculturist, engineer, politician, development administrator, public health specialist, and economist. Although their various attitudes are somewhat caricatured in the following sections, it is hoped that important elements of truth are revealed in this distorted picture.

Agricultural viewpoint

Irrigation facilitates crop growth in an unfavourable environment. Where it is arid, irrigation is essential. For example, in Sind in West Pakistan average rainfall as 12–30 cm but potential evapotranspiration is more than 150 cm. Without irrigation there would be only a desert flora.

Irrigation is also desirable where rainfall distribution over the year does not coincide with demands for crop water. The soil has a limited capacity to act as a reservoir for moisture. If dry periods

Notes and references for this chapter begin on page 266.

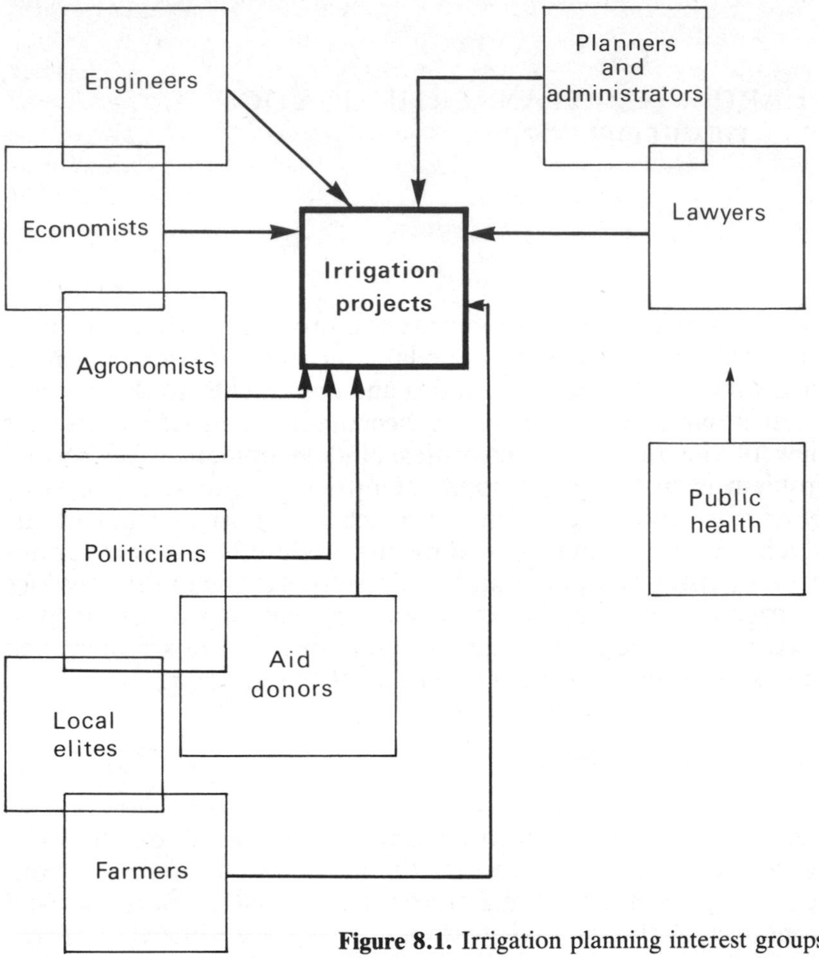

Figure 8.1. Irrigation planning interest groups.

exceed about 30 days, there is likely to be a yield response to irrigation. Irrigation is also useful where rainfall is uncertain. Supplementary irrigation gives security and may extend the season for high-value crops.

Agriculturists are all too often interested in the technology of maximum production, usually measured by the highest yield per acre. The new technology that has been provided largely by plant breeders has sharpened their focus. Assured supplies of water are necessary to the varieties with a high-yield potential. Unfortunately, the agriculturists' preoccupation with yield may limit a broader vision that includes the opportunity costs of obtaining the yield and the social consequences of the uneven access within a country to irrigation facilities.

Agriculturists also support irrigation in order to enable farmers to cultivate crops such as rice and sugar-cane which have high water requirements and low tolerance for moisture stress. They are interested in soil improvement by drainage and crop protection from flooding. Drainage is often needed after irrigation disturbs the sub-soil water balance. Drainage can be regarded as irrigation in reverse and the economic principles of planning irrigation applied to drainage investment.

Engineering viewpoint

For a variety of reasons, some of which are discussed below, engineers are often criticized by economists for their excessively technological outlook. Before joining in this criticism, the authors would like to acknowledge that by no means all engineers can be thus labelled. Furthermore, the majority of irrigation schemes are well conceived, designed, and executed, and, though operation and maintenance procedures generally leave much to be desired, most projects perform adequately from a technical viewpoint. It is not the work of engineers that is sometimes questionable but their point of view.

Empty, arid land promotes a conditioned urge among many irrigation engineers, and they seek to discover means for its development. If a river runs nearby or groundwater underlies the area, it is almost impossible to restrain them. Close proximity of arid land and 'wasted' water provides, in their view, both necessary and sufficient conditions for irrigation development. Furthermore, sometimes engineers are too concerned with structures—large interesting structures, particularly if their fees are proportional to construction costs.

Professional engineering standards are high and rigorously enforced. This is excellent, but it can lead the designer to be excessively cautious and inflexible. Many of the professional standards used in poor countries were devised for application in wealthy countries where the resource endowments are different in whole and in parts. In developed countries, all labour is scarce, and capital relatively plentiful and cheap. In contrast, in developing countries unskilled labour is plentiful, and capital scarce and expensive. It is therefore highly unlikely that the least-cost solutions, the engineering standards and codes of practice established for rich countries will be appropriate in poor countries.

Engineers may argue, with some justification, that certain standards are not variable (for example, safety factors on a dam). This is often so but there are alternative ways to attain these standards. For example, there is often a choice between labour-intensive or capital-intensive construction methods.

There is more flexibility in design standards in areas where safety is not such an obvious consideration. It is worth noting that it is sometimes possible to save scarce capital resources with a lower-cost construction method and to accept that this will mean higher recurrent costs later—for example, wooden buildings using local materials may be erected rather than concrete buildings using imported cement. An important element in deciding on economic choice in this situation is discount rate policy. Other things being equal, a higher discount rate will favour saving on capital and accepting higher recurrent costs. However, the current practice of most aid agencies in providing capital assistance (normally foreign exchange only) tends to provide economic support for the capital-intensive solutions.

It is possible to find engineering designs which take no account of the timing of the costs and benefits. For example, a recent authoritative text on canal linings and canal seepage, which purports to 'summarize the factors to be taken into account when considering a canal lining programme', makes no mention of discounting[1] although the lining costs occur at one point in time and benefits arise later, over a number of years. The economics of canal lining are extremely sensitive to manipulations of discount rates and canal lining will be favoured with a low discount rate.

The engineering profession sometimes uses standard assumptions to overcome problems of difficult analysis. For instance, it may decide to use in design a water availability figure related to (say) the 1-in-5 dry year, or to design a flood protection project for the 1-in-100-year flood because of precedents. No consideration is made of the marginal benefit and marginal cost for accepting a 1-in-6-year, 1-in-4-year drought, or a 1-in-60-year or 1-in-200-year flood. It is always incumbent upon the engineer to justify his assumptions and criteria.

Political viewpoint

Politicians are often attacked by the 'desert bloom' syndrome that afflicts engineers. The symptoms of this affliction are worst in the true desert regions where water is clearly identifiable as the basis for life and regard for it is built into local customs, culture, and religion. In Christianity, Judaism, and Islam there are texts that give special praise to 'desert bloom' activities. Politicians may get their inspiration from these. Irrigation is also one of the more rapid and dramatic forms of public investment that a politician can promote. Dams can be named after him, canal opening or well pumping can be filmed and photographed, and within a few weeks the sand is green and farmers have a new way of life. Promotion of irrigation will benefit all constituents in the way that a school or health-centre cannot, and with luck it can be conceived and construction completed or well under way within a politician's five-year term of office.

These characteristics of irrigation development, plus the fact that they are generally non-controversial projects (unlike prisons or military aid) generally commend irrigation to aid donors. This fact is appreciated by recipient governments, and there is a risk of excessive requests for aid with excess capacity being subsequently created. This may account, in part, for the poor performance of many projects.

Analysts should not be too hard or cynical about the role of the politician. In many circumstances development goals can appropriately be met through irrigation. There are areas such as Kigezi in western Uganda where water control in swamp areas is one of the few means open for alleviating high population densities. In other arid areas, such as Pakistan, irrigated agriculture is one of the few means open for creating job opportunities for the growing population. Irrigation may also contribute to lessening expenditures on famine relief in pastoral areas such as in the Sahel zone in Africa. It may even be rational and reasonable for politicians to use irrigation settlements as a means for demarcating international boundaries, as have Israel and Kenya. Well planned and excuted irrigation is one of the most promising ways of reaping the benefits of modern agricultural technology, and as such it deserves political support.

Viewpoint of development administration

Development administrators and professional planners understand that it is in rural areas that a solution must be sought for the pressing problems of unemployment and food supplies. Many of the major problems of rural development stem from the lack of control over both the producers and the production processes. As previously noted, the new agronomic technology requires precise control. For example, the new seeds require adherence to a definite irrigation and husbandry schedule, which can be achieved either by informed, receptive farmers or by discipline. Often the former conditions do not obtain; hence the administrators have to rely upon discipline. Irrigation facilitates discipline—

it is not a coincidence that the original organized societies and early 'hydraulic societies' were almost all characterized by their reliance upon the manipulation of water resources.[2]

The potential for social control by irrigation schemes has been used by the successful National Irrigation Board of Kenya, notably at Mwea,[3] to obtain high levels of technical performance from inexperienced peasant farmers. Water supply and, from this, all aspects of production are controlled by project management. Discipline is maintained by enforcing strict rules, which includes the possibility of ejection from a holding, while maintaining high incomes (over $400/farmer/yr, more than three times off-scheme income).

In other schemes, the control element has made possible profitable links with processing industries which require long production runs and consistently high-quality produce, e.g. in Kenya the harvesting season for canning crops such as pineapple is extended and quality is improved by overhead irrigation systems. In Bangladesh, irrigation of tea enables estates to extend the plucking season, and increases yields to give an estimated 15 per cent rate of return to investment.[4]

Development planners may also view irrigation as one part of multi-purpose schemes; they may promote, for instance, projects linking irrigation and hydro-power or irrigation and drinking-water supply.

Where settlement of farmers in previously unused land is part of an irrigation project, planners often give excessive stress to social infrastructure—housing, roads, schools, and so on, and to relatively high farm size and high levels of minimum income. The desire to avoid the creation of a 'rural slum' is understandable, but it results in high capital costs per family—i.e. in Kenya $\$_{74}$ 1200 to 3800 per family for 1·62-hectare holdings—and perpetuation of poor living conditions in the overcrowded rainfed areas from where the settlers are drawn.

Public health viewpoint

The medical or public health perspective contains less economic rationale than those previously discussed, perhaps because decisions on human life are involved. Nevertheless, the provision of public water supplies or the establishment of health safeguards on irrigation schemes requires commitment of scarce resources. Therefore economic decisions are involved and are taken by default if not design.

In attempting to apply any form of analysis to investment

decisions which will in time benefit public health, a major problem arises in that many data elements in any model are unknown. For instance, medical experts appear to know remarkably little in quantitative terms about acquired immunity to disease, about the overall levels of morbidity in rural areas, about the effects of maternal fitness on fertility, or about the fundamental causes of infant mortality in rural areas. Given these disciplinary inadequacies, the economic analysis which requires such information as an input may be considered to be premature.[5]

In these circumstances, the public health administrator relies upon professional judgement in determining priorities. These judgements, like those of the engineers, are normally conservative. The economist must accept this working procedure but can assume responsibility for indicating the impact and opportunity costs of the resources thereby committed.

The health aspects of irrigation development are typically neglected, and impose social costs upon communities dependent upon irrigation. Tropical diseases have been classified according to their mode of transmission by Bradley and modified by Feachem.[6] This classification is shown in table 8.1. It is categories 3 and 4 which cause the main problems in irrigation schemes and reservoirs.[7]

Many of these diseases are debilitating and chronic, and they will obviously affect labour productivity. Preventive or control measures can be taken, but, despite the logic of the argument for such measures, success in their adoption in irrigation planning is as rare as (or more rare than) in other areas of preventive medicine.

Irrigation development often means rapid building of townships to house construction workers and sometimes displaced farmers. Preventive public health measures (sanitary facilities, water supplies, immunization, and prophylaxis) can help at this stage. In the operation phase there is more surface and perennial water, which, with a denser population, may lead to unhealthy conditions. In a previous chapter, drainage was discussed in terms of the benefits to crop production; but adequate drainage is also important for control of insect and snail populations which effect the incidence of several severe diseases. It is therefore important that the multi-disciplinary team typically involved in planning new projects and rehabilitating old ones includes an epidemiologist.

On the Mwea Rice Irrigation Scheme in Kenya, a chemical snail-killing preparation was successfully applied (it killed all snails 16 km down-stream of the point of application) at an

Table 8.1. Classification of water-related diseases and prevention strategy.

Category	*Example*
1. Faecal-oral (water-borne or water-washed)	
(a) low-infective dose	Cholera
(b) high-infective dose	Bacillary dysentery, ascariasis
2. Water-washed	
(a) skin and eye infections	Trachoma, scabies, leprosy, yaws
(b) other	Louse-borne fevers
3. Water-based	
(a) penetrating skin	Schistosomiasis
(b) ingested	Guinea worm, paragoniniasis
4. Water-related insect vectors	
(a) biting near water	Sleeping sickness
(b) breeding in water	Malaria, onchoceriasis
Transmission mechanism	*Preventive strategy*
Water-borne	Improve water quality; and prevent casual use of other unimproved sources.
Water-washed	Improve water quantity; improve water accessibility; and improve hygiene.
Water-based	Decrease need for water contact; control snail populations; and improve quality.
Water-related insect vector	Improve surface water management; destroy breeding sites of insects; and decrease need to visit breeding sites.

Source: Bradley, *op. cit.*

annual cost of $\$_{74}$ 1·00 per hectare or $\$_{74}$ 1·60 per family. Although it was not possible to demonstrate conclusively that this is a cost-effective measure, it is clear that snail control is a precondition for bilharziasis control and that $1/ha is a relatively small part of production costs. In 1976 malaria prophylaxis was administered to all staff, tenants, and their families as well as bilharziasis control. The costs were $\$_{74}$ 1·2/ha for malaria prophylaxis and $\$_{75}$ 1·7/ha for bilharziasis control. This is an enlightened and probably an economic measure but only possible on the tightly controlled settlement schemes of the Kenya pattern which are producing profitable crops.[8] (It should be noted that the profitability of the Kenyan irrigation schemes is to some extent illusory. The irrigation authorities have a monopoly of production of rice and domestic production is protected by import controls.)

Farmer involvement

The first draft of fig. 8.1 excluded the farmers, although we have been using this diagram, or something like it, for a decade.[9] Why is this? Could there lie behind this admission, which on reflection reveals a widespread attitude, a partial explanation for poor performance of plans?

It is easy to understand why farmers have been by-passed in the past. Firstly, agricultural information is often held in low regard by planners. It is only relatively recently that professional agriculturalists have been involved in irrigation planning and still their role is limited to providing the model cropping patterns which form the basis for irrigation design and which may, or may not, be adopted by farmers. If professional agriculturalists were excluded then it is to be expected that farmers would also. Secondly, many irrigation projects were for ‘empty’ areas or for farmers totally new to irrigation. Thirdly, as farmers are a relatively large and heterogeneous group, a few representative farmers might be hard to identify. Fourthly, the generally low level of education does not facilitate effective communication with planners. By and large, it is our experience that most planners have a paternalistic view of their role and an unhealthy regard for the possible value of farmer participation in the planning process. Failure to appreciate and be sensitive to the socio-political realities of the social environment into which irrigation technology is to be introduced is a sure recipe for problems later. Agriculture has special problems not encountered in industry and the way in which these problems influence decision making at the farm level is best understood by farmers.

Economic viewpoint

Economists regard irrigation planning primarily as an exercise in resource allocation. Decisions regarding investment should be made in the light of information on the economic, financial, and social contributions of the scheme as well as technical, political, administrative, and legal criteria. Irrigation choices should not be determined simply by engineering or agronomic reasoning, nor by the whim or hunch of a particular individual. If such procedures are followed, then the role of the economist is to indicate the costs (in terms of opportunity cost) of the decision.

Water 'fundamentalism', a term for the elevation of water issues to special status, is a real danger because of the implication that water is, in a special sense, basic to life, and a gift from God. This is essentially false. Only 2–3 litres per day are basic to life, and within limits it does not have to be clean or accessible, or sufficient for irrigation. Whereas rain and river water might reasonably be regarded as a gift from God, canals, pipeline structures, and the rest are clearly man-made and not gifts. Thus the economist's litany is that irrigation facilities require the use of a variety of scarce resources with alternative competing demands. In selecting the optimal use of these resources the economists must look hard at objectives and alternatives; he (or she) must forecast future events and weight benefits and costs arising at different points in time and to different groups; he must take note of the market price signals and take account of distortions; he must study the whole irrigation system and make judgements on the appropriate institutional framework; and finally the economist must provide feedback information to planners and operating agencies. Interests of economists in project planning will be discussed at greater length later in this chapter.

MULTIPLE OBJECTIVES

Historically, the main objective of an irrigation project was to generate a profitable agriculture that would enable the initiators—governments or individuals—to obtain a reasonable return on the capital invested. How today's planners must yearn for such a simple goal! Contemporary public planners are set multiple objectives, some of which are complementary, some of which conflicting. In the planning phase, financial aspects are less important but returns to the overall economy—social returns—are central. Finance has, of course, to be found, but the direct revenues do not have to exceed costs. Added to this there are often goals related to income redistribution, employment creation, regional development, national self-sufficiency, and nutrition.

There are three main approaches to the problems of planning for multiple objectives. In the first procedure, the advantages and disadvantages of each project in relation to the various objectives can be listed in tabular form to show decision makers the consequence of specific actions. The trade-offs inherent in a choice are then clear. If this approach is adopted for all irrigation projects (and other agricultural projects), the implicit weights attached by decision makers to various goals become clear and this can be accounted for in design. For example, if great weight is being given to employment creation among unskilled labour, then labour-intensive designs can be favoured, even at a slightly greater financial cost.

The second approach to planning for multiple objectives is the constraint approach. Here the contribution to one objective is over-riding; other objectives must be satisfied at prescribed minimum levels of performance, i.e. a maximum economic rate of return might be sought, subject to a specified level of income redistribution to a defined low-income category of farmers or landless labourers in the project area.

The third approach is the multiple objective function approach. Here the diversity of objectives is incorporated into a single objective function with specified weights attached to each objective. This is the most complex procedure which requires a clear statement of objectives, the relative weighting of each objective and a method of conversion of non-additive benefits into a common unit or numeraire. It is tantamount to the specification of a social welfare function, and it has found much favour in recent writings on project appraisal.[10] Operationally, there are still many problems. The experience of the authors in many countries suggests, given the present planning capacity of most irrigation agencies and the lack of economic sophistication of many decision makers, that this method is not generally appropriate. Many planners and decision makers are baffled by the mechanics of the procedures. A procedure is not to be recommended if the workings of the criteria are not evident to the decision makers. For this reason the first tabular or decision matrix approach, illustrated in table 8.2, is likely to be preferable at present for irrigation planning.[11]

This decision matrix satisfied what Chambers[12] describes as the growing needs for appraisal methods for small projects with a poverty orientation. He contends that selection procedures should be simple, open to inspection, intelligible, and sparing of the scarce skills and time needed in preparation. No single technique in economic appraisal can corner all bases for choice. In project selection, economic objectives are only one of several criteria that must be taken into account. Given the present pressures upon the planning and decision-making system, it is contended that advanced project appraisal methods cannot yet be considered satisfactory. It is contended that the specialist who is not an economist could, by scanning a decision matrix, consider simultaneously a wide range of relevant issues and thus make

Table 8.2. Illustrative layout of economic decision matrix.

	Criteria[2]					
		Financial internal rate of return[3]				
Project alternatives[1]	*Economic internal rate of return*[3]	*To farmers*	*To government*	*Jobs created/ $1000 investment*	*Proportion of project income to poorest 20% of population %*	*Location in priority development area (Yes or No)*
1	*	*	*	*	*	*
2	*	*	*	*	*	*
3	*	*	*	*	*	*
4	*	*	*	*	*	*
.	.	.	.	.	.	.
.	.	.	.	.	.	.
.	.	.	.	.	.	.
n	*	*	*	*	*	*

Source: Carruthers and Clayton, *op. cit.*

1. These could be projects of different scale, timing, location *or* projects of a different nature. However, the validity of comparisons is lessened if very unlike projects are compared, e.g. an irrigation project and a hospital.

2. These criteria are examples only. Specific criteria would be set according to (i) objectives of policy, (ii) constraints existing in the economy, and (iii) the time to full development.

3. Risk is more important in agriculture, therefore the plan should include a risk assessment, the forecast range and frequency distribution of rates of return.

better selection decisions than with the more sophisticated procedures. The first criterion in economics would require consideration of the more important elements of social cost benefit analysis. Transfer payments, sunk costs, and double counting of benefits have to be avoided; and discounting (at 10–15 per cent) and some simplified form of shadow pricing must be adopted. The simple, pragmatic approach, as advocated by Gittinger,[13] with prices adjusted in the right direction to the approximate order of magnitude, appears to the authors to be correct. A shadow exchange rate should, where possible, be supplied by the Treasury. World market prices should be used for inputs and outputs. A shadow wage rate halfway between the marginal product of labour in agriculture and the money wage would suffice for most purposes. This approach would, in the authors' view, release irrigation planners to concentrate upon technical and economic issues as outlined below:

forecasts of production with and without the project;
quality of produce;
inter-seasonal and intra-seasonal fluctuations in supplies;
market structures;
the effect of differential inflation on factor costs;
management systems;
farmer response to price changes;
financial forecasts and revenue policy formulation.

These items appear from ex-post studies to be the most important and neglected aspects of agricultural projects.

CONCERNS OF ECONOMICS

In the previous section, some matters of importance to economic planners were outlined. Irrigation project planning economists have their own special concerns. Amongst them are the consideration of alternatives, projections of future events, decisions involving choice of timing, price distortions, of the importance of a system approach and of institutions. These will be considered below.

Alternatives

Economists are interested in alternatives and in marginal adjustments (such as larger or smaller schemes) in more or less water per hectare, in earlier or later construction. In interdisciplinary irrigation, economists for planning groups often have to question the technical assumptions of their colleagues. The unpopularity of some economists resulting from such boat-rocking activities and the increased work load demanded in considering more than one means to achieve the project objectives is part of the penalty of improved economic planning. It is a penalty that should be amply covered by the cost savings and higher benefits made possible by improved planning.

Prediction

Economists must be able to predict future events in order to judge the merits of alternative irrigation proposals. All irrigation projects promise a stream of future benefits, and the level of each benefit and the uncertainty associated with this estimate must be assessed. Agronomists, and to a lesser extent engineers, are often wary about projecting technical parameters, such as crop yields, for ten years or more ahead. It might be thought that getting technicians to estimate uncertainty would be even more difficult. However, practical experience indicates that if the range and probability of occurrence of a parameter are sought, the task is in fact simplified.[14]

A variety of data and an understanding of the existing situation and trends in the level of major variables are prerequisites for agricultural forecasting. Data collection for agriculture is an expensive and exacting task. Agricultural units are normally numerous, small in size, scattered across large areas, sometimes with a wide range of resource endowments, with physical output levels subject to weather, pests and diseases, and with value of output subject to price fluctuations. Economists are very conscious of the costs of obtaining improved data, and are interested in the problems of planning with inadequate information.

Choice of timing

Jam today and jam tomorrow are certainly not the same thing. Economists are concerned with the weight put upon consumption in different periods of time. The weighting decision involves the choice of a discount rate and application of this to costs and benefits. The rate selected is a very important determinant of least-cost technical solutions. A high discount rate favours the substitution of capital costs by recurrent costs, longer-term projects, and phased construction. Irrigation planning economists should seek guidance from the central planning unit or financing authorities on the rate to be applied. In public projects, the choice is usually between two rates, one relatively high, taking account of social-opportunity cost; the other, a relatively low rate, favouring social time preference. The former stresses forgone opportunities in the public sector, the latter the need not to give too little consideration to the welfare of future generations.[15]

Price distortions

In most developing countries the market prices for the inputs and outputs from irrigation do not reflect the true value to the economy. There are many reasons for this, but the basic explanation lies in the methods used by government to regulate the economy (e.g. tariffs, import duties, quotas, export taxes, income taxes, minimum wage legislation, unemployment, unequal distribution of wealth, or segregated and deficient capital markets). The consequence of government interference is compounded by poor communication systems: a peasant farmer may not know that a 50 per cent rate of interest for credit is higher than that charged in urban areas, and, even if he is aware, he may not be able to do anything about it.

The effect of market imperfections is that the prices of key factors of production and product prices do not reflect their real value, or their opportunity cost, to the economy. Economists therefore select and substitute in appraisal 'shadow prices' for market prices.[16] (A major defect of the design sequence is that engineers seldom consider or find it practically possible to use shadow prices in selecting least-cost designs.[17]) The most important shadow-priced items are usually foreign exchange (undervalued), product market prices (under- or over-valued), unskilled labour (over-valued), skilled labour (under-valued). Use of unadjusted market prices for irrigation normally leads to excessively capital-intensive construction and agricultural production methods.

The extent to which government policy can obscure the returns to irrigation is shown in an extreme fashion by the case of rice in Sri Lanka. In 1971 each eligible person was given free each week a measure of rice (nearly one kilo) and another measure at a highly subsidized rate. This subsidy consumed about 20 per cent of annual expenditure of the State. Half of the foreign-exchange earnings of the country are expended upon food. Any assessment of the relatively successful increased rice production programme should be made using world market prices rather than one or other of the local prices.[18]

System approach

For each project, economists are interested in the feasibility of the over-all system of irrigated agriculture to be adopted or already in effect. For example, in project planning, checks are typically made on availability of labour, power, and of water for each half month; and constraints are placed upon cropping patterns by consideration of minimum subsistence food requirements. Thus food grains may take precedence over more profitable cash crops. The requirements of each crop enterprise are identified, and from this the most profitable crop combination can be selected that is consistent with the resources available for exploitation. The procedure adopted for determining the design cropping pattern is discussed later and illustrated in fig. 8.4.

Institutions

Economists can identify policy instruments and actions desirable for the attainment of policy goals, but these actions will be implemented through the existing institutions. Institutional aspects of development which must be considered in prediction of irrigation development include the land tenure pattern, credit supplies, agricultural research and extension, and the marketing system. In recent years, partly because of the excitement and promise of the Green Revolution, the institutional reforms necessary for successful irrigation have been neglected. The World Bank[19] puts stress upon needed reforms to wasteful water management and maintenance systems. Inappropriate forms of local water control exist where large and influential landlords obtain a disproportionate and timely share of water resources and yet use them inefficiently. The major and steadily growing problem of fragmentation of land holdings in some areas precludes land levelling and shaping, efficient extent and form of groundwater exploitation, reduction of land wastage in boundaries, and elimination of wasteful travelling time between plots.

Ruttan (1977) contends that technical change in agriculture is endogenously generated but that institutional change is not occurring to keep up with the technical potential. He states that:

> (the) developing world is still trying to cope with the debris of non-viable institutional innovations: with extension services with no capacity to extend knowledge or little knowledge to extend; cooperatives that serve to channel resources to village elites; price stabilization policies that have the effect of amplifying commodity price fluctuations; and rural

> development programmes that are incapable of expanding the resources available to rural people.
>
> . . . unless social science research can generate new knowledge leading to viable institutional innovation and more effective institutional performance, the potential productivity growth made possible by scientific and technical innovation will be under utilized.[20]

We would concur with this perspective. Our experience of planning water use in agriculture suggests that political economy coupled with scientific insight is the only suitable frame-work for evaluation of development that is likely to lead to an improved performance in the vital rural sector.

ACTIVITIES IN PROJECT PLANNING

Irrigation project planning should be a complement to macro- and sector-planning activities. The major links are shown in fig. 8.2 with the sector study and project planning activities elaborated in fig. 8.3.[21]

Project planning consists of a set of activities which include the definition of precise objectives, the selection of criteria or rules for assessing the contribution of alternative means of satisfying objectives, the collection of information to assist in applying criteria, and the final choice of the detailed form of the project and its priority with the national programme.

A detailed schematic representation of the sequence of these activities in cropping pattern and capacity design is presented in fig. 8.4. This is only one way of proceeding. The basic problem in practical project planning is that each specialist needs the completed work of a colleague in order to carry out his specialist task. The engineer needs to know the water requirements to design his canals or wells, the crop physiologist needs to know what crops may be sown to work out requirements, the agriculturist needs to know the value of products to determine cropping patterns, and the economist needs to know the availability of water to determine the profitable crop mix.

Ideally, project planning should be an iterative procedure, with the initial rough assumptions being modified at each iteration. To a limited extent this happens in planning groups. However, the authors know from experience that all too often the 'tentative' cropping pattern becomes firm and final merely with the passage of time because so many dependent calculations have been made on the original assumptions.

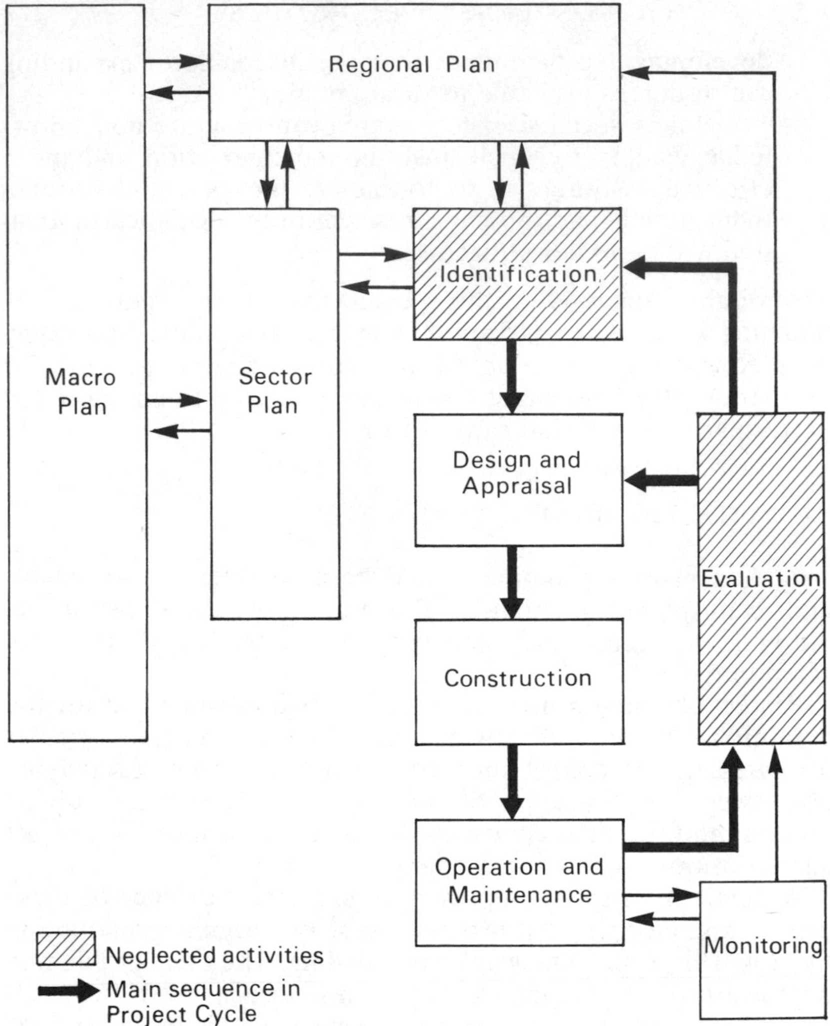

Figure 8.2. Schematic representation of irrigation planning activities.

In fig. 8.4 it is assumed that the objectives and the intensity dilemma discussed in chapter 1 are resolved in advance. A tentative cropping pattern is then selected on the basis of *a priori* technical and economic arguments or on the basis of the most profitable proven combination of enterprises. In the first attempt at designing a cropping pattern, a simple budgeting approach is advocated. However, the problem is one of simultaneously finding the optimum combination of various possible crop and livestock activities subject to a set of resource constraints. A

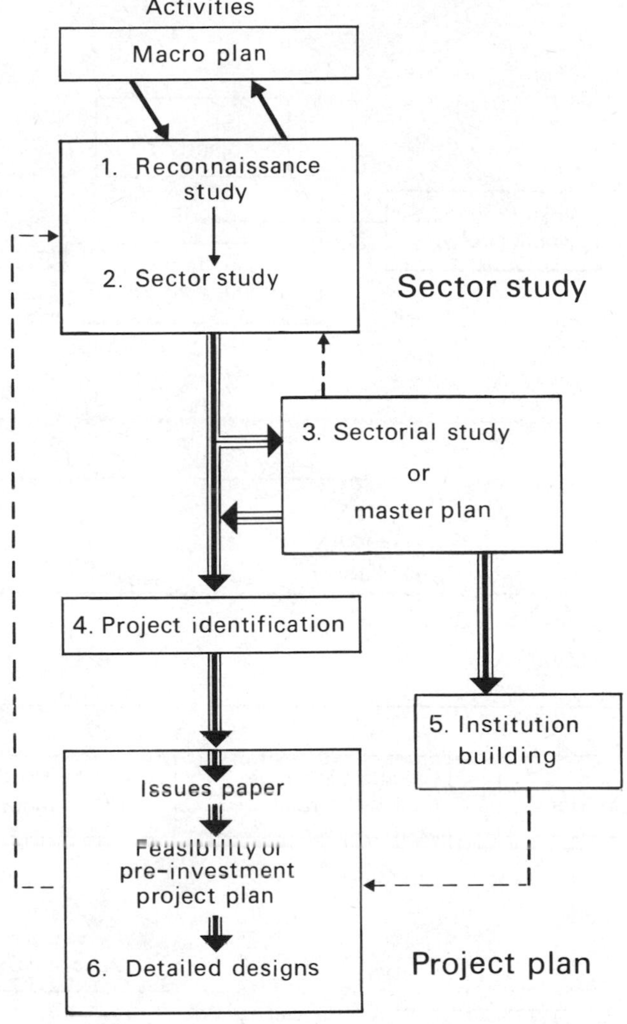

Figure 8.3. Planning procedures in the water sector.

Functions:

1. Inform governments of objectives and procedures for sector reviews and other activities.
2. Data assembly, constraint analysis performance evaluation, indicative, strategic planning. Can usefully be aggregated from regional studies.
3. Long range, more detailed and integrated sector plans with appraisal of alternative development strategies.
4. Detailed investigations to evaluate alternatives which may or may not result in projects.
5. Institution strengthening or building to cope with planning training and operation of water projects.
6. Technical, institutional economic and financial assessment of project followed by comprehensive and systematic review by financing agency.
7. Detailed designs drawings, tender documents, bills of quantities, cost estimates and contracts.

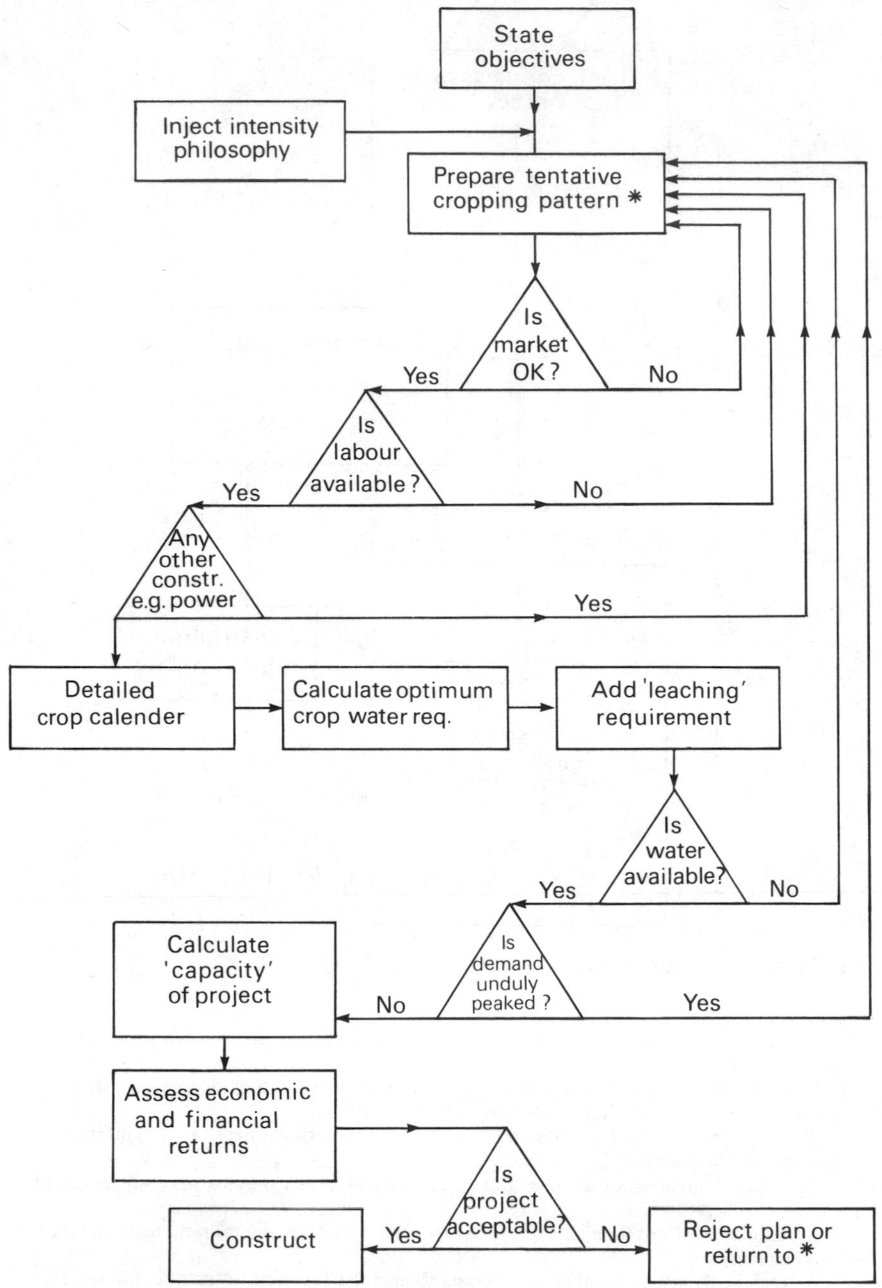

Figure 8.4. Cropping pattern determination.

problem with n activities and m resources, one or more of which is limiting, is a classic problem for a linear programming (*LP*) technique.

Several attempts have been made to incorporate LP into project design and, although the outcomes are extremely promising, the process is not yet fully operational. One of the main areas of difficulty is the specification of water availability to allow for soil moisture storage and compensatory growth in subsequent periods. Other areas of difficulty include data problems, access to computers, inherent limitations of the technique (linearity assumptions, divisibility (e.g. 1·7 cows), activity independence, deterministic rather than stochastic data coefficients, short-term static solutions) and the obvious problems of abstracting from the complex social, cultural,[22] and economic environment using any mathematical model. However, although these evident limitations indicate caution in use of the LP technique, no agricultural economist can have even the slightest exposure to it and fail to make a giant conceptual leap in the viewing of problems as a simultaneous set of opportunities and resource constraints in matrix form. Thus the most valuable contribution of LP is not in full field use, which is generally not feasible because of data limitations, but in changing ways of thinking from single-crop enterprises separately optimized to multiple, interdependent cropping systems optimized for the farm or region.[23]

Once the tentative cropping pattern is established the market for the likely output is then tested in two stages. Firstly, the general market for the products or the market structure is assessed, followed by an estimation of the present size of the market and its apparent growth, as well as present and future factors influencing demand, including income, prices, availability of substitutes, and the influence of public tastes. Secondly, the place within this market structure of the products of the project have to be assessed, including the relative costs of production and transportation.

In the market analysis the existence of pecuniary externalities stemming from the project should be checked and the impact included in the analysis. This may include increase in prices for inputs or complements as a result of project demand or a fall in price of outputs of the project or goods competing for the project output. Technological externalities which may be positive (demonstration effects and training) or negative (pollution, congestion, waterlogging or increased bilharziasis) should also be recognized and the effects incorporated into the overall cost-benefit assessment. These external effects will be par-

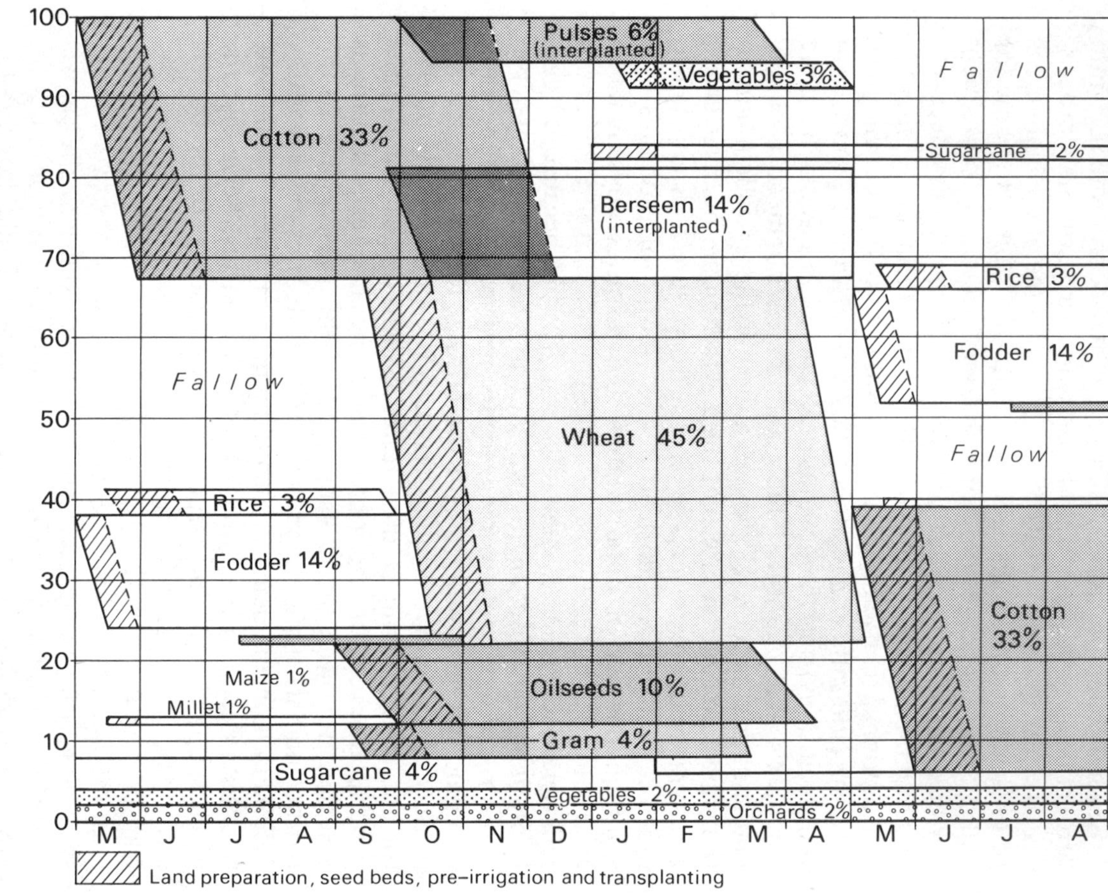
Area (ha)
or % land
100
90
80
70
60
50
40
30
20
10
0
Cotton 33%
Pulses 6%
(interplanted)
Vegetables 3%
Fallow
Sugarcane 2%
Berseem 14%
(interplanted)
Rice 3%
Fodder 14%
Fallow
Wheat 45%
Fallow
Rice 3%
Fodder 14%
Cotton
33%
Maize 1%
Oilseeds 10%
Millet 1%
Gram 4%
Sugarcane 4%
Vegetables 2%
Orchards 2%
M J J A S O N D J F M A M J J A
Land preparation, seed beds, pre-irrigation and transplanting

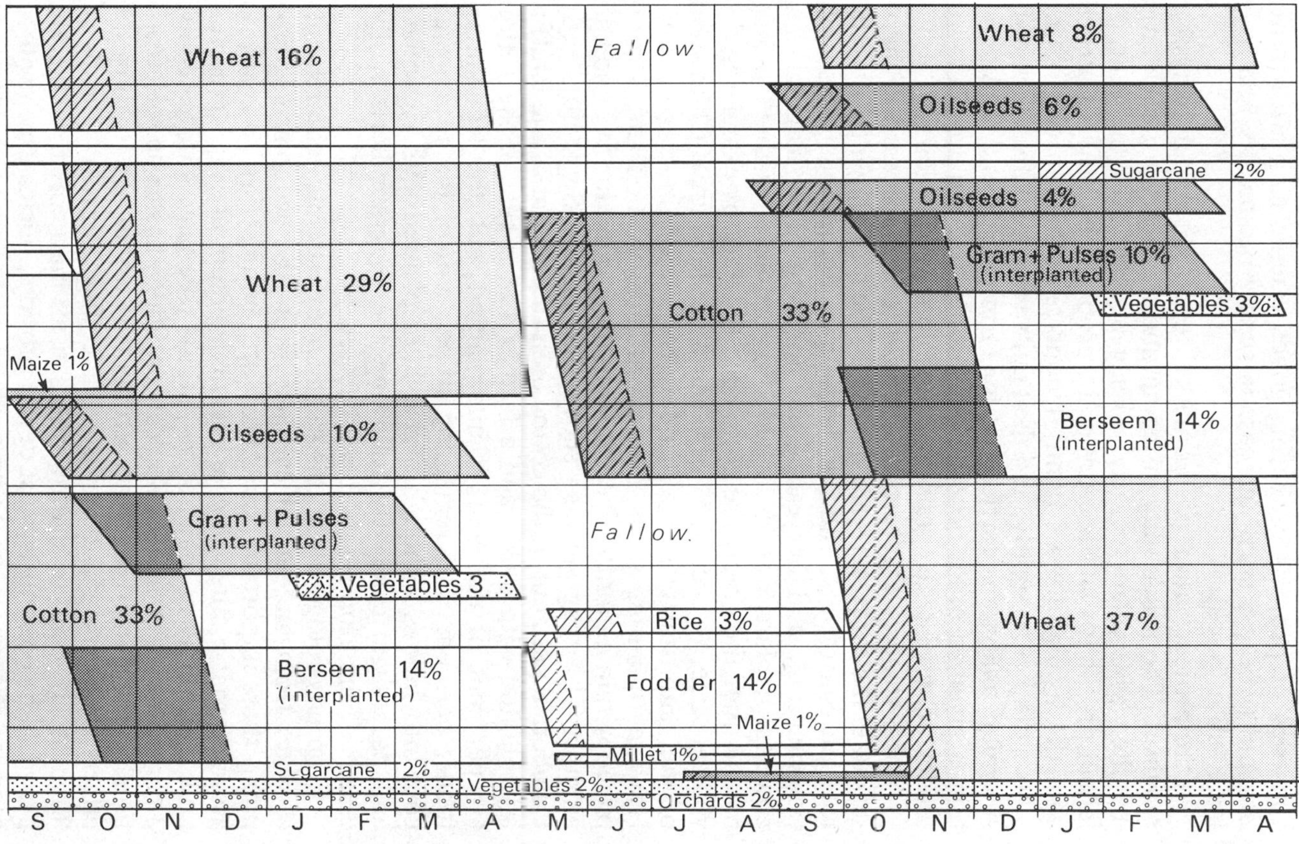

Figure 8.5. Cropping pattern and rotation display.

ticularly important where there are large projects in small countries or projects producing goods when facing an inelastic demand (where a small increase in supply produces a relatively large fall in price).

If the project output passes market tests then labour analysis is required. The availability of labour and the needs for each crop are then assembled for, say, 15-day periods over the year. A labour profile for all the crops in the proposed pattern is matched with the availability of family labour plus the capacity of the labour market to supply, and the family to employ, hired labour. A similar analysis is carried out for farm power, and consideration is given to other possible constraints, i.e. requirements for fodder or subsistence grain or constraints on rotation.

At this stage a detailed crop calendar can be prepared. If this is set out in the form shown in figure 8.5, the cropping intensity, the rotation constraints, and the duration of the periods of sowing, transplanting, and harvesting are apparent. This diagram is a powerful explanatory visual aid. For example, in the top left-hand corner the enterprise cotton can be seen, by examining the vertical scale, to occupy 33 per cent of the land area. Cultivation of the land starts in May and is completed in two months. However, planting starts a month after cultivation and is completed within a month. The crop is in the ground for five and half months. The diagram also shows that in late September/early October 18 per cent of the cotton area (6 per cent of the land) is interplanted with pulses and a further 42 per cent of the cotton (14 per cent of the land) is interplanted with the leguminous fodder berseem. The remaining 40 per cent of the cotton area goes into fallow, although part is later planted with vegetables and part sugar-cane. By following the diagram horizontally, details of the crop rotation can be seen.

Crop physiologists use this calendar to estimate requirements of water from meteorological data. From the economist's viewpoint, three important points are worth noting. First, it is always assumed that plants will consume their potential evapotranspiration, which may or may not be optimal (see page 43). Secondly, a particular set of meteorological conditions has to be selected—usually, the 1-in-10 dry year. Economists would like to know the effects on costs and benefits of selecting alternative levels of 'dryness'. If social or shadow prices differ markedly from market prices then economists must devise subsidy policies which will be required if farmers are to be encouraged to grow crops that are socially profitable (say, providing valuable foreign exchange) using socially desirable means (say, labour intensive cultivation methods rather than tractors). Thirdly, the cropping pattern selected is based upon current market price and forecast yield levels. Nobody assumes that relative prices and profits will remain constant, and market prices may not equate with social prices. Cropping patterns will be modified in line with changing opportunities. In view of this and the fact that the major determinant of water requirement are energy inputs and stage of growth, it is considered that this aspect of planning could be greatly simplified with 'model' crops grown. Massive fieldwork efforts by crop physiologists to measure the different water requirements of, say, mustard and wheat which may or may not be grown in practice indicate a misunderstanding of the problem and a misplaced effort.

Irrigation project design in arid areas generally includes a leaching component to ensure that sufficient water passes through the soil to depress salt toxicity. This is added as a percentage of half-monthly water requirements, and canals and tubewells are designed to supply this level. Although it is necessary to leach the soil regularly—say, every one or two years—there is no need to apply it in even doses throughout the year. As every farmer appreciates that the value of water depends upon the time it is available, it is unlikely that he will add between 6 and 20 per cent extra water[24] when it has highest value to leach the soil. Such a farmer will logically grow 6 to 20 per cent more crop, and leach the soil in the months when water has low, zero, or even negative value—in the Lower Indus in Pakistan, for example, after rainfall or during August, February, or March.

This set of derived demands for water is then compared with the available water in each ten to fifteen day period. Normally, there will not be an even level of irrigation demand over the year, especially where there are clear rainy seasons or strong seasonal temperature differences, as occurs with projects distant from the Equator. Some peaking of demand is acceptable, but financial problems are likely to increase as the irrigation delivery system is idle for a higher proportion of the year. In the case of high-value crops, an irrigation season of just three months (that is, 25 per cent or less utilization) may be profitable.[25]

The capacity of the project can now be established. From this and the crop requirements the engineering capital and recurrent costs can be derived at market and social (shadow) prices. At the same time the agricultural returns using private (market) and social prices can be derived.

Three major analyses are then carried out. Firstly, an economic analysis to show the rate of return or cost–benefit ratio; secondly, an estimate of financial returns to farmers; and thirdly, after a consideration of the farmers' taxable capacity and feasible rating systems, an estimate of financial returns to the responsible agency.

Economic analysis is carried out using shadow prices, by determining the incremental value of production over what would have happened in the absence of the project. In addition to adjustment of inputs and outputs to shadow prices, care should be taken to avoid double-counting costs and benefits; increased value of land and increased value of production after irrigation represent essentially the same phenomena and are not separate benefits. Furthermore, transfer payments within the economy (land purchase for construction, taxes and water-rates, grants and subsidies, import and export duties) should be deducted before the economic calculation as they neither add nor deduct from national product, though there may be important income distribution effects. There is an increasing emphasis and special study of the incidence of benefit and analysis of the outcomes giving particular weight to benefits obtained by low-income groups (see Squire and van der Tak, *op. cit.*).

It should be noted that the net value added is the difference between value added with and without the project, not value added before and after the project. This difference can be significant as, for example, in the normal situation when some improvement could have been expected in agriculture even in the absence of a project. In these circumstances a 'before' and 'after' test would exaggerate the benefits attributable to the project. In some cases using a 'before' and 'after' test rather than 'with' and 'without' test will lead to an under-estimation of the net benefits. For example, in a situation where deterioration can be expected in the absence of a project—say, from waterlogging or salinity—the with and without project test will yield higher benefits. It has to be admitted that forecasting the without project situation, which normally shows agricultural improvement, is even more hazardous than making 'with project' forecasts.

In order to forecast the rate of uptake of irrigation, the level of yields, and relative changes in price (but not inflation), the range of possible outcomes of technical parameters and the probability of their occurrence must be established. Various ways of deriving this technical data and of assessing the risk of wrong estimates from experts are available.[26] Broadly speaking, it is recommended that analysts retain, as long as is practicable, the expert's specification of a range of possible outcomes and the subjective probabilities of each outcome occurring. This increases the computation load, but it improves the validity of the prediction.

Table 8.3. Sensitivity tests on return of irrigation in Rahad, Sudan.

Test	*Internal rate of return (%)*
Project design	18·8
Cotton yields minus 25 per cent	14·0
Cotton prices minus 25 per cent	13·5
Cotton and groundnut prices minus 25 per cent	12·0
Cotton yields and prices minus 25 per cent	8·5
All yields and prices minus 25 per cent	6·0
Labour costs $1·67 per day	18·0
Labour costs $2·5 per day	16·0
Labour costs excluded	21·5
Machinery and material costs plus 25 per cent	17·5

Source: A. M. Ibrahim and J. Berkoff, 'An economic evaluation of the Rahad Irrigation Project' R47 Q33 II *International Commission for Irrigation and Drainage, 10th Congress Proceedings,* Athens, 1978.

Sensitivity analysis of the effects on returns of changes in key assumptions can be an important part of the economic analysis. This is illustrated in table 8.3 which shows the outcome of a sensitivity analysis on the Rahad Irrigation Project in Sudan. In this case the sensitivity analysis indicates that the crucial assumptions are cotton yields and prices. Therefore, the project analysts can check the technical assumptions which underpinned the yield and price forecasts. In the operation phase priority can be given to ensuring that cotton yields, and as far as is possible price forecasts, are actually achieved.

Particular emphasis needs to be given to the iterative nature of this planning procedure. A change in any assumption emerging from an analysis of, say, crop opportunities, or water availability at a particular time of the year, will require a change in the system and the procedure has to be restarted anew or an amended cropping pattern devised.

In assessing crop production benefits not only increased yields but also increases in quality, improvements in location or timing of production, decreases in transport and other production costs, avoidance of losses, and so on, are all sources of benefit for the economic analysis.

FINANCE ASPECTS AND REPAYMENT IN PROJECT PLANNING CAPACITY

In assessing financial returns to producers, market prices are used. From this analysis can be seen the financial incentive to production in new irrigation systems. This can be compared to the effort required of the farmer to achieve the forecast level of production which, when weighed against the reduced risk and minimum acceptable income levels, will facilitate judgements on payment capacity. The experience and economic principles that would guide such judgements are set out in chapter 7.

Once economic and financial tests are completed, the desirability of the project can be assessed. At this stage, decision makers will apply other criteria, including technical, administrative, legal, social, and political criteria. These criteria do not have equal weight. Therefore it is perfectly possible that a project which is economically attractive will be turned down, although we must admit that the converse is more often true.

EVALUATION OF DATA IN PROJECT PLANNING

As mentioned at the beginning of this chapter, project planning involves a logical sequence of activities. These start with identification of problems to be encountered, and continue with their appraisal. Then follow the design, construction, operation, and maintenance of the scheme. The final phase in this sequence is evaluation. This is based on the sum of recorded experience of operating schemes.

Insufficient attention is given to evaluation. The efficiency of the whole sequence is dependent upon a feedback loop from evaluation to the early phases. Project planning is thus an iterative procedure. We have already seen that recent evaluations indicate a number of testable propositions regarding irrigation schemes. These include the likelihood that irrigation costs are under-estimated; that construction time is under-estimated; that uptake is slower than forecast; that economic returns are lower; that financial returns are minimal; and that electric power uptake is more rapid than forecast. Designs are usually too sophisticated, and unpredictable exogenous factors dominate success or failure. Poor operation and maintenance is a key constraint.

The lessons of operating experience are very largely ignored in planning. The design of dams, wells, and structures is a specialized task and the practitioners seldom have either operating experience or access to analytical studies of factors leading to success or failure in previous projects. The planning process which is characterized by the consultant-client-contractor relationship, and very limited public discussion of either objectives or means, does not assist widespread availability and learning from past mistakes.

Ex-post evaluation should be a systematic review procedure. There is the danger of an evaluation study being a vague, historical or descriptive treatise, of little value for prescription and containing few concrete lessons. A major methodological issue inhibiting a systematic approach relates to choice of criteria or preference tests. Obvious criteria will be of a technical, economic and financial nature but other goals may be set concerned with employment, regional impact, nutrition, and self-sufficiency which will also require tests of achievement. A decade ago, economic efficiency was regarded as a valid proxy measure for social welfare goals. Its main defect arose from the need to give a premium to savings and a higher priority to individuals, groups or regions by-passed by the economic growth process. This led to the advocacy of the more complex procedures for assessing the impact of alternative projects upon inter-temporal and inter-personal welfare. Criteria set in relation to various goals are not inherently equal. Whilst technical feasibility may be considered a necessary condition for development, over-riding all other criteria, various standards of technical performance above some minimum levels may be optimal. Evaluation studies should examine several aspects of technical performance. The relative weight given in evaluation to technical, economic, financial, legal, administrative and other aspects of project performance in order to establish the degree of success cannot be predetermined, but they are weighted implicitly or explicitly in the design and appraisal stages.

The neglect of ex-post evaluation is a major reason for the repetition of planning mistakes, as well as poor operating performance. However, the authors do not anticipate much change in this situation because rigorous evaluation is generally not in the short-term interest of the operating authorities or the aid donors (Carruthers and Clayton, *op. cit.* 1977).

WATER MANAGEMENT

Deficiencies in management systems are an increasingly common diagnosis for the failure of irrigation projects to reach their potential levels of productivity. The commonplace failure to achieve adequate and timely water supplies by farmers at the tail of canals and because of illegal actions of those at the head is also attributed to lax management. This diagnosis begs the question, to some extent, since it is extremely difficult to define what is meant by good management. In the farm economics literature returns due to 'management' are generally derived by establishing the productivity of land, labour, and capital, and attributing the residual returns to the management factor. This is clearly an unsatisfactory procedure. In the case of irrigation projects a somewhat analogous procedure seems to have been adopted. The argument that management problems limit potential economic returns to irrigation has been arrived at latterly when other technological causes have been largely eliminated. (Unfortunately the proposition is not testable because management performance in low-income countries is not necessarily linked to project performance.) Exogenous factors, outside the control of management, impinge in numerous ways upon scheme productivity. Some of these will be outside the control of the country (e.g. world inflation, shortages of materials, slow delivery of orders for equipment). Solutions to many domestically generated problems are also outside the control of management. These relate to major matters, such as the failure of other agencies to deliver electricity on time or complete a road, to distortions in pricing policies and other market defects; or to relative minor matters, such as late delivery of fertilizers, failure of buses to deliver workers on time or the imperfect functioning of telephones. The joint effect of such features is to excuse management in a genuine way from responsibility for achieving potential productivity levels. It makes the task of assessing management performance extremely difficult and imprecise.

Considerable progress has recently been made in research into the study of the management process in irrigation. Pioneering research by Wade and Bottrall has shifted the emphasis away from technical and operational reforms to the neglected aspects of man-management.[27] Bottrall, following a global literature review, lists the favoured remedies for rehabilitating malfunctioning irrigation systems.[28]

(1) Technical improvements:
  (a) remodelling/rehabilitation of main system (additional control structures, etc.);
  (b) improved physical layout at tertiary level ('on-farm development').
(2) Organizational changes:
  (a) formation of single co-ordinating agency at project level;
  (b) strengthened agricultural extension;
  (c) creation of formal 'water users' associations.

(3) Economic/financial measures:
 (a) higher water rates;
 (b) larger budget allocations for operation and maintenance.

(4) Operational procedures:
 (a) rotational irrigation at the tertiary level;
 (b) improved techniques of water application at farm level.

He argues persuasively and quotes supporting studies to show that improved man-management, with reform of the objectives, procedures, functions, and responsibility of the irrigation bureaucracy, is a grossly neglected area. He notes that this topic is not on the list of favoured solutions. Chambers supports this contention and illustrates the problem by reference to the 1969 three volume FAO report for the 360,000 ha Mahaweli Ganga Project in Sri Lanka which had less than one page on the structure of irrigation management and not a word on operational procedures.[29]

Often it is the management information flow which is ill-considered. Few monthly or annual reporting systems meet the minimum requirements for effective management control. Yet, paradoxically perhaps, management is typically overloaded with repetitive reports, demands for information for outside agencies, excessive detail in some reports, poorly designed forms, delays in communicating information, missing data, and in some instances wrong or doctored data.[30] Seldom are irrigation managers trained to set up management systems and consultants tend to get involved with grand structural change such as setting up new ministries or parastatal bodies rather than the details of field personnel management or detail design of simplified monitoring routines (i.e., forms, filing systems, etc.).

There is a growing awareness among practitioners and academic observers of irrigation development of disappointing performance of existing schemes, of the pitfalls of new projects, and of the potential for achieving production and equity objectives through improving operation and management of existing schemes. In 1976 a group of 35 specialists from a dozen countries concluded[31] that the potential gains from improved operation and management would be realized only if there was:

(1). Evaluation and feedback of current performance including political and organizational factors which impinge upon operational decisions, management of staff who control and issue the water and incentives to farmers to economize on water use.

(2). Prestige and resources for operation and maintenance activities more in line with the more glamorous than that accorded to the tasks of planning, design and construction.

(3). Political support for necessary but unpopular measures such as close control of surface water allocations and possibly groundwater abstractions, for payment of water charges, for firm measures against bribery or threats to field staff, and for prompt and impartial prosecution of infringements to the legal regulatory framework.

(4). Improved technical training and extension of the principles of controlled water use.

(5). Development of small irrigation groups for improved farmer-official contacts relating to such matters as local operation and maintenance, and to technical training.

(6). Modernization, intensification and expansion of infrastructure, particularly rehabilitation of medium-sized or small-scale systems and at watercourse or farm level, which has been a much neglected activity despite the substantial benefits which have, in some circumstances, been demonstrated.

(7). Political priority for effective improvement of operation and maintenance and for the often low status irrigation ministry or department.

Many of these assertions, although emanating from an authoritative source, are little more than untested hypotheses. However, there is sufficient empirical evidence to indicate that they are credible propositions with important policy implications when verified in a particular environment. Management of operating schemes is clearly an area worthy of further research.

CHAPTER 9 SUMMARY AND CONCLUSIONS

ECONOMIC THEORY AND IRRIGATION

In this book the authors have identified the main economic principles that underlie the economics of irrigation. However, following the early tradition of economics, they have attempted to test the validity of the various concepts and theory by mustering empirical evidence on as wide a range of experience as was feasible. This was a salutary exercise which, given the wide range of economic performance that was established, may appear to detract from the apparent elegance of formal theory. However, the fairly broad perspective presented in this book has confirmed that the principles of opportunity cost, effective demand, marginal adjustment, and economies of scale (to give some important examples) do provide valuable insights which, for all the defects of the real world, reveal a certain symmetry of experience which can assist in policy formulation and execution.

It is fortunate that it can be concluded that basic theory has some validity because of the increasing global importance of irrigation. Four to five thousand years ago, the civilized world in North Africa and the Middle East, in Indo-Pakistan and the Far East, and in Central America, was dependent upon irrigated agriculture for its growth and prosperity. Today the challenge in supplying food is much greater, but once again the greatest opportunities lie in irrigated agriculture.

IRRIGATION AND TECHNOLOGY

Modern agricultural technology has created opportunities for crop production which markedly increase the return to investment in irrigation. However, controlled water supply produces

Notes and references for this chapter begin on p. 268.

the very highest yield response when used in conjunction with other inputs such as fertilizer, crop protection from pest and disease, high quality in the preparation of land, and timely weeding and harvesting.

Has modern technology placed agriculturists in a technological bind where the adoption of one part of the technology (say, improved seeds) requires the adoption of other complex, expensive innovations, often dependent on fossil oil? Does the failure to provide one part of the package (as must be almost inevitable in low-income countries) preclude maximum success over-all?

To some extent this is so. Some years ago Herdt and Mellor[1] demonstrated that for irrigated rice the total yield per hectare and the marginal response to nitrogen fertilizers was greater in the United States than India at any given level of nitrogen. In addition, in the United States the response was positive at a much higher level of nitrogen application. This does not imply that there would be higher marginal response at present levels of fertilizer (or pesticide) application in rich countries than in poor countries. Indeed, the marginal response, in kilogrammes of grain, at present levels of application of fertilizer and water is said by some observers to be much higher in poor countries.[2]

This may not be the case. The result of thousands of experiments in Asia summarized by FAO in *The response of rice to fertilizer* (FAO, Rome, 1966) showed a marginal return, at low input levels, of 10 kg unmilled rice/kg nitrogen element. Even after the injection of genetically superior varieties into Asian agriculture in the late 1960s and early 1970s Barker reports that the rule of thumb that 1 kg N/ha produces 10 kg rough rice is still valid in selected villages he studied.[3] However, in the low application villages the marginal response was higher (12–35 kg per kg nitrogen). There was no evidence that it would be worthwhile increasing fertilizer inputs in any villages given present environmental and management conditions, although experimental yields were at least 2 tons per hectare greater on average.

The low levels of application are partly the response to higher input prices and partly the rational response to the absence of other complementary factors of production in the desired amount and at the right time. For example, even for many farmers on irrigation schemes, the supply of water is irregular because of inadequate maintenance of facilities or poor regulation of water distribution. In peasant agriculture the large number and wide variety of proximate constraints result in the threshold level for economic return to fertilizer or water soon being reached. In economic jargon the profit surface with respect to fertilizer or water rapidly becomes flat.

The picture is not totally depressing. It is now established that high yielding varieties (HYV) perform better in physical and financial terms than local varieties in adverse as well as in good conditions.[4] As a consequence it cannot be claimed that, if part of the modern agronomic package is missing, such as reliable irrigation, no worthwhile developments can be expected. The latest releases of new rice and wheat cereal varieties are generally much more robust than HYV or local varieties that were available in the early 1960s. Many productive developments can be expected to be worthwhile without a complete package of inputs. Nevertheless, the availability of one part of the package increases the marginal return to other elements. Modern technological developments increase the options for productive investments in agriculture and thereby greatly complicate the process of appraisal of proposed irrigation investments.

WATER REQUIREMENTS OF CROPS

Arbitrary judgements have hitherto been made about the supposed water requirements of different crops. If some crops are grown on unusually porous soils, important differences may arise in water loss by seepage. But otherwise there should be no differences between water requirements of different crops per day of growing season, apart from those arising from certain minor factors, such as the crop not completely covering the soil at certain stages of its growth, and the glossiness or roughness of the leaves. For the amount of water which a crop transpires is that required to remove enough heat to keep its temperature down to a bearable level, which in turn depends on the external physical factors of solar radiation, wind, heat storage in the soil, and so on. A heavy crop also requires no more water than a light one (pot experiments on this point can be misleading, because they do not reproduce the thermal conditions of the field, where evaporation from one plant helps to cool its neighbours). Also, contrary to what is widely believed, crops heavily fertilized with nitrogen require no more water than light crops, and indeed are better able to withstand temporary water shortages. HYV cereal crops require no more water than traditional varieties. Indeed, to the extent that they are short-duration crops, their total water requirement would be less.

From the above follows the important economic conclusion

that (other things being equal) it is always worthwhile to concentrate irrigation water on the highest-valued and heaviest-growing crops. The level of water use which will result in maximum evapotranspiration, whereas it is likely to produce maximum yield per unit of land, will not produce maximum returns per units of water. Water is all too often the scarce factor which should be spread to those areas where highest response (net of other costs) can be obtained.

Plants have a surprising degree of flexibility in tolerating and producing at levels of soil moisture below that necessary to ensure maximum evapotranspiration. The beliefs that each plant has a unique water requirement which must be met, or that the capacity of the system must satisfy potential evapotranspiration, have been shown to be fallacious.

It is unfortunately not yet possible to construct a production function showing the yield from water supplies at various levels below the optimum. Over-watering not only fails to give any additional yield, but also may, if persisted in, have very serious consequences in raising the water-table to plant root level, thus making the land uncultivable, as in large areas of West Pakistan. This can, however, be cured, at considerable expense, by adequate drainage.

WATER RESOURCES AND DEMANDS MADE ON THEM

There is no danger that the world will 'run out' of water. Certainly, supplies are finite, and there is a growing claim upon limited fresh water supplies. The population of towns and cities is expanding rapidly as a consequence not only of immigration from rural areas but also of improved public health; this latter results, in part, from a more liberal use in urban areas of a freely available water supply. Increases in urban incomes have the effect of raising levels of *per capita* consumption of water, as house connections, multiple outlets in the house, and even garden watering, become the norm. Curbing this increase in urban consumption is difficult because rationing is impracticable and, to a great extent, water use is independent of the price charged. Industry in cities is notoriously wasteful of water because the technology was devised for locations where supplies were ample and cheap. Furthermore, the legal framework within which industry developed allowed industry to avoid the costs of cleaning the waste water before it was returned to the water courses. In consequence, much industrial use is wasteful and has resulted in widespread pollution.

Since the recent oil price rise, urban demands are stretching out into the countryside in the form of demands for hydro-electric schemes. Hydro schemes may indirectly assist rural use by regulating stream flows, but this is by no means certain. Normally, up-stream abstractions for agricultural use are impeded following hydro development, and availability of water downstream is conditioned by industrial and urban demands for electricity, which may or may not coincide with water requirements for agriculture.

In any situation of potential conflict, urban demands for water will generally take priority over rural needs because the existing urban supplies are typically polluted, alternative sources of water are distant, the health risk, including epidemic disease, is high, and the economic gains are generally greater than in rural areas. In addition, in cities, the financial returns are likely to be sufficient to allow operation of the supply as a public utility.

The consequence of these increasing urban demands is that the rural people see their streams and rivers depleted and polluted. This is particularly unfortunate because it is occurring at a time when opportunities for water use in the countryside are increasing both in number and value. Although urban demands have priority, it is irrigation which really consumes large quantities of water.

A convenient unit for measuring national water resources is that which would serve to irrigate a million hectares to a depth of 1 metre annually ($10^{10}\,m^3/yr$). Total stream flow in the United States is 187 units, of which agriculture's net use is about 15 per cent, with about 25–30 per cent required to keep the rivers flushed at a barely tolerable level of cleanliness. This should, on the face of it, leave an ample margin for domestic and individual uses. It is, however, least available in the times and places at which it is most required. In some of the drier western states, it has been found necessary to construct water systems at rapidly mounting costs; and it appears that some irrigated agriculture will have to be abandoned.

India has a stream flow of 167 units, of which it is hoped eventually to use 34·7 for irrigation, about the maximum proportion that could ever be practicable, in view of the great irregularity of the flow of the river. With the Nile and the Indus, however, in spite of the irregularity of flow, it is hoped eventually to use three-quarters of the flow after the construction of some costly works.

The available flow of water in rivers depends, of course, on the rainfall, of which, however, a considerable amount may evaporate before it reaches the rivers; and unfortunately it is in the areas where water is needed most that the greatest proportion is lost in this way. Run-off (river flows) appears to be related to rainfall by a cubic function—proportionate run-off being much higher where rainfall is highest, with different coefficients to the function for different types of topography.

IMPORTANCE OF ECONOMIC ANALYSIS IN IRRIGATION SCHEMES

If irrigation is to make the maximum use of the scarce fresh water supplies, considerably greater efforts must be made to improve management of schemes and efficiency of distribution. Canal lining, control of water-table, and drainage are all areas of decision which relate to this goal. Economic analysis can assist in making choices.

Great care must be taken to ensure that the economic models used in such decisions are fairly specified and that data is correctly assembled to facilitate proper estimation of relationships. For example, it is clear that the farmers' understanding of the water response function (which is very different from the usual conception of the irrigation scheme designer) may lead to spreading of water and thus to levels of water application per hectare that are too low to ensure that, in arid zones, soil salinity does not build up to harmful levels. The authors have taken care in this book to treat this important consideration by advocating that leaching requirements are specified but that leaching water is applied when there is low (or zero) water value in use.

COSTING AND IRRIGATION

Development of groundwater promises to be an important area for future investment. A chapter is included here in which are discussed the important issues that arise in attempting to integrate this resource with other forms of water development. Groundwater is typically a property resource of open access and common to all and, although private development is possible and in some ways desirable, public regulation is necessary if the old overgrazing 'tragedy of the commons' of nineteenth-century Britain is not to be repeated. For, with an open access common property resource, freedom of use brings ruin to all. Already there are severe problems (social costs) associated with excessive groundwater pumping in some parts of the world which demonstrate the truth of this adage.

Primitive methods of raising water from wells by both man- and ox-power prove to be barely remunerative taking into consideration high Indian prices for grain, low Indian wages, and the fact that the water-table is near the surface. However, simple mechanical pumps prove highly remunerative under these circumstances. Pumping costs are lowest if electric power is available; diesel pumps incur higher costs for both fuel and maintenance.

Irrigation costs have increased as the low-cost, easily developed sites have been exploited and the best soils cultivated. They have, of course, also increased as a consequence of world inflation, but, as inflation affects both costs and benefits, this problem may be much more apparent than real. However, there is some evidence that inflation of irrigation construction costs has been offset to a significant extent because engineering design and construction has been an area of cost-reducing technological advance. This results in a decrease in real costs per unit of water delivered, which increases the relative attractiveness of irrigation compared with alternative investments. In particular, there have been considerable advances in methods of structure and canal design using computers, in groundwater exploration, well-drilling, and screens, and the technology concerned with earth moving and concrete manufacture. The availability of cheap plastic fittings has increased the efficiency of application of several delivery systems, thus increasing the availability of water and reducing the cost per unit of water delivered to the crop.

The costs of constructing dams show returns to scale. Thus the cost is less per unit of stored water for the largest dams, where a great deal more water is stored per unit volume of dam wall. Unless, however, the water can be distributed down a river system, the savings of a large dam per unit cost are mainly offset by the additional costs incurred for the distribution network (and sometimes for land levelling).

In recent years a more cautious attitude to large dams has been adopted. This is not simply because the low-cost options are already developed. There is an increasing administrative and political concern about stability and safety of large structures, about the large-scale and irreversible nature of the commitment, and about the risks of environmental or social disruption. And there is growing evidence that improving the existing infrastructure, by modernizing outdated equipment and improving organization and management, is a much more cost-effective form of investment.

Costs of water for irrigation vary greatly according to whether it is drawn from the flow of streams, pumped from wells of varying depth, or obtained by constructing dams and distribution channels. Spray and trickle irrigation, whose techniques have recently improved, may be no more costly than furrow irrigation.

In the chapter on irrigation costs it was shown that there are complex interrelationships between economies of scale in water storage developments and increasing costs as the obvious low-cost projects are undertaken and long distribution systems are developed. This is complicated by new cost-reducing technical developments. No general prescription can be made except perhaps to examine costs carefully and occasionally to re-examine projects which were previously regarded as uneconomic. Given the low system efficiency in many schemes, it is clear that rehabilitation of early constructed low-cost schemes is likely to be more cost-effective than building new schemes in unfavourable areas.

IRRIGATION AND RETURNS

In commercial agriculture, economic returns to irrigation can be measured in c/m^3, for comparison with costs. In some countries, however, agricultural production is primarily for subsistence rather than for sale. Alternatively government subsidy or tax policy may result in the prices of crops being quite out of line with those prevailing in world markets. In these cases, therefore, it is often better to measure the returns per cubic metre of irrigation water supplied in terms of kilogrammes of wheat equivalent, other agricultural products being expressed in terms of wheat on the ratios prevailing in the world markets.

The authors note that there is often a large gap between predicted returns and realized returns. Irrigation does reduce the risk of crop failure, and the increment can be substantial. However realizing the potential requires not only a good irrigation supply but also a range of complementary agricultural and institutional support (for example, improved agricultural research and extension). This is why it seems that high returns to irrigation occur in rich countries. Efficient operation of irrigation schemes appear to be a consequence, rather than a cause, of development.

In making overall judgements on the economics of irrigation, it is important not to define the debate too narrowly. The impact of irrigation on irrigation farmers or operating agencies should not be considered to the exclusion of the impact on non-irrigated farm producers (whose market position may be affected) and consumers of farm produce who may benefit from lower prices.

CHARGES FOR THE WATER SUPPLY

In improving irrigation performance, the water-rate might appear to be a useful instrument of policy. Indeed, it can be used to give the desirable signals to operators of the real costs of irrigation. Alternatively, low prices, or even free supplies, can be provided to promote use. In fact, an economic price is seldom charged. Furthermore, there appears to be little improvement in rich countries over the practice in poor countries. Benefit rather than a cost basis for pricing has attractions in poor countries, where low-income farmers are close to subsistence and where failure to achieve potential yields may be no fault of the farmer. However, public revenue at the disposal of government is perhaps the most scarce resource in the majority of developing countries. As irrigation farmers are the easily identifiable direct beneficiaries of public investment, it is reasonable to expect some revenuc. In inflationary periods, water rates are seldom adjusted regularly in line with changes in operating costs. A lack of direct revenue may contribute to poor maintenance of facilities. Crop input and product prices are often distorted by public policy and the gains from irrigation activity are spread widely through society so that it is often difficult to assess whether schemes are 'subsidized' even though direct revenue levels are low. A study of the literature reveals that the field of water-tariffs has an over-developed economic literature but it is an area very much neglected in terms of political action.

Water-rates are a policy instrument that could be used to help improve income distribution patterns as well as scheme operating efficiency. Equity criteria are given increasing weight in political/economic planning. Some critics of the seed–fertilizer–irrigation package of technology contend that it has been produced or promoted by elite groups to support their own interests; or that it has been taken over by, and reinforced the power of, elite groups; or that it has itself created new elites in rural areas to the detriment of the welfare of the poorest groups. Any contemporary assessment of the economics of irrigation cannot readily ignore these issues because the policy implications, if any of these contentions are true, are quite evident and important. In the authors' view the technology of irrigation is not intrinsically maldistributive. It is true that certain attributes of irrigation technology can be an asset to a strong government seeking to promote policies of rural development (and to a selfish landlord seeking to maintain his economic and social status). We have the lessons of history in the ancient irrigation civilizations to show that this is so. But control of irrigation cannot turn a weak government into a strong government. Judgements on whether, on balance, irrigation creates more problems or harmful effects than benefits are likely to be value judgements. A review of commentaries on the Green Revolution suggests that the outcome from such a consideration may hinge as much on whether the writer considers the term 'exploitation' to represent a virtue or a vice as on the apparent facts.

Despite the apparent plethora of writings on pricing theory, in comparison to the field of planning it is an academic desert. Project planning manuals appear to be promoted by almost every aid agency. It is not too cynical to note that cost-benefit analysis, as presently practised, had its origins in the water sector in the 1930s in the United States where various government agencies were anxious to promote worthwhile public projects which did not appear sufficiently attractive with formal financial tests. At that time, market tests did not give the correct economic signals. Similar conditions exist in developing countries today. The elaborate, sometimes almost metaphysical, procedures devised to cope with market price distortions and multiple objectives are typically applied most rigorously in the water resource sector.

In searching for the reasons for disappointing performance belated attention is being given to the management of water distribution at the bulk (canal) level and at the field level. Such studies present theoretical and practical problems. Management performance cannot be judged by results in low-income countries because of the large influence and complexity of exogenous factors such as inflation, poor weather, or unreliable supply of materials. Nevertheless, there is increasing recognition of the importance of effective management and the institutions which support this service to the outcome of irrigation projects. It is contended that to be successful, irrigation development must include not only the hardware but also the software of services, institutions, and attitudes. It is widely believed by practitioners and academic researchers in the irrigation field that priority should generally be given to improving existing schemes and not to creating new schemes. Existing schemes require improved management systems which implies better information systems; more prestige and resources for operation and maintenance;

extension and training in water management; better organization at field level; rehabilitation and modernization of existing infrastructure; and political support for irrigation policy including necessary but often unpopular reforms (e.g. water rationing or water-rate increases).

In the authors' view, the appropriate irrigation planning and management procedures should be less concerned with the full rigour of modern cost-benefit analysis and other management techniques and more involved with study of the salient facts and testing of options, using criteria derived from insights obtained from basic socio-economic concepts. This book has attempted to illustrate how this approach could be used to improve decision-making among agriculturists, engineers, medical specialists, administrators, and, above all, politicians, in this increasingly important area of technology.

APPENDIX. USEFUL CONVERSION FACTORS

Professional irrigation engineers in English-speaking countries have been accustomed to thinking in terms of acres, acre-feet and cusecs, and may be inconvenienced by having to translate these units into 0·405 hectares, 1234 cubic metres, or 895,000 cubic metres per year, respectively. But this inconvenience will be outweighed by the far greater ease with which problems can be grasped when stated in metric units. A million cubic metres of irrigation water, when applied to a depth of one metre (quite a usual seasonal depth) will irrigate exactly one square kilometre, or a metre of rainfall on the same area will represent another million cubic metres. British units are not so conveniently inter-related. It is true that, fortuitously, 1 cusec = 1 acre-inch per hour or 2 acre-feet per day. The unit in which all quantities of water are expressed is the cubic metre (for which m^3 is the customary abbreviation). Conversion factors that may be useful are set out here together with details of the various national currencies at different periods of time.

The following conversion factors may be useful:

1 cent/cubic metre = $12·34/acre-foot
1 cusec/square mile/year=34·5 centimetres depth/year
1 kilogramme/hectare = 0·89 lb/acre

A more complete list is presented in table A.1.

Table A.2 gives, for early 1974, the purchasing power, in current dollar units, (abbreviation as $ (dollar) and c) of the currencies of some leading countries using irrigation.

For agricultural products, on the other hand, United States prices are often well out of line with those prevailing elsewhere. The natural unit in this case is the kilogramme of wheat, all other agricultural products being expressed as wheat in terms of the relative prices prevailing in the local markets. In assessing benefits this economic conversion factor has been used—not the relative calorific value, which would unduly depreciate some high-protein and other agricultural products.

Table A.1. Useful conversion factors.

Conversion factors for area

1 acre	=	4840 yd^2
	=	4046·9 m^2
	=	0·4047 hectare
	=	4·047 × 10^{-3} km^2
1 sq. mile	=	640 acres
	=	2·590 × 10^6 m^2
	=	259·0 hectares
	=	2·590 km^2
1 sq. metre	=	1·196 yd^2
	=	2·471 × 10^{-4} acre
	=	1 × 10^{-4} hectares
	=	1 × 10^{-6} km^2
1 hectare	=	2·471 acres
	=	3·861 × 10^{-3} mi^2
	=	1 × 10^4 m^2
	=	0·01 km^2
1 sq. kilometre	=	1·196 × 10^6 yd^2
	=	247·1 acres
	=	0·3861 mi^2
	=	1 × 10^6 m^2
	=	100 hectares

Conversion factors for volume

1 cubic yard	=	0·7646 m^3
	=	6·198 × 10^{-4} acre-foot
1 acre-foot	=	1234 m^3
1 litre	=	1 × 10^{-3} m^3
1 cubic metre	=	1000 litres
1 Imperial gallon	=	4·546 litres
1 US gallon	=	3·785 litres

Conversion factors for flow rate

1 cubic foot per second	=	1·98 AF/day
	=	28·3 l/sec.
	=	0·0283 m^3/sec.
	=	2450 m^3/day
1 acre-foot per day	=	0·505 cfs
	=	14·2 l/sec.
	=	0·0142 m^3/sec.
	=	1230 m^3/day
1 litre per second	=	0·0353 cfs
	=	0·0703 AF/day
	=	0·001 m^3/sec.
	=	86·4 m^3/day

Conversion factors for flow rate (cont.)

1 cubic metre per second = 35·3 cfs
= 70·0 AF/day
= 1000 l/sec.
= $8{\cdot}64 \times 10^4$ m³/day

1 cubic metre per day = $4{\cdot}09 \times 10^{-4}$ cfs
= $8{\cdot}11 \times 10^{-4}$ AF/day
= 0·0116 l/sec.

Conversion factors for weight

1 pound = 0·4536 kg

1 short ton = 2000 lb

1 long ton = 2240 lb
= 1016·0 kg
= 1·016 m ton

1 metric ton = 1000 kg
= 0·9842 long ton

1 kilogram = 2·205 lb

Table A.2. Official exchange rates, 1974, 1975 and 1976. National currency per US dollar (mid-point rates), end of period.

Country or area	*Unit*	*1974*	*1975*	*1976*
Afghanistan	Afghani	45·0	45·0	45·0
Algeria	Dinar	4·0	4·1	4·4
Australia	Dollar	0·8	0·8	0·9
Bangladesh	Taka	8·1	14·8	15·0
China	Yuan	2·0	2·0	2·0
Egypt	Pound	0·4	0·4	0·4
Ethiopia	Dollar	2·1	2·1	2·1
France	Franc	4·4	4·5	5·0
Germany, Fed. Rep.	D. Mark	2·4	2·6	2·4
Ghana	Cedi	1·2	1·2	1·2
Greece	Drachma	30·0	35·7	37·0
Guyana	Dollar	2·2	2·6	2·6
India	Rupee	8·1	8·9	8·9
Indonesia	Rupiah	415·0	415·0	415·0
Iran	Rial	67·6	69·3	70·6
Iraq	Dinar	0·3	0·3	0·3
Israel	Pound	6·0	7·1	8·9
Italy	Lira	649·4	683·6	875·0
Japan	Yen	301·0	305·2	292·0
Jordan	Dinar	0·3	0·3	0·3
Kenya	Shilling	7·1	8·3	8·3
Lebanon	Pound	2·3	—	2·9
Libyan Arab Rep.	Dinar	0·3	0·3	0·3
Morocco	Dirham	4·2	4·2	4·5

Country or area	*Unit*	*1974*	*1975*	*1976*
Nepal	Rupee	10·6	12·5	12·5
Nigeria	Naira	0·6	0·6	0·6
Pakistan	Rupee	9·9	9·9	9·9
Peru	Sol	38·7	45·0	69·4
Portugal	Escudo	24·6	27·5	31·6
Saudi Arabia	Riyal	3·6	3·5	3·5
Somalia	Shilling	6·3	6·3	6·3
South Africa	Rand	0·7	0·9	0·9
Spain	Peseta	56·1	60·0	68·3
Sri Lanka	Rupee	6·7	7·7	8·8
Sudan	Pound	0·3	0·4	0·35
Swaziland	Lilangeni	0·7	0·9	0·9
Switzerland	Franc	2·5	2·6	2·5
Syrian Arab Rep	Pound	3·7	3·7	4·0
Thailand	Baht	20·4	20·4	20·4
Tunisia	Dinar	0·4	0·4	0·4
Turkey	Lira	14·0	15·2	16·7
United Kingdom	Pound	0·4	0·5	0·6
USSR	Rouble	0·8	0·8	0·6
Un. R. of Tanzania	Shilling	7·1	8·3	8·4

Source: UN Statistical Yearbook, 1977, (New York, 1978).

NOTES AND REFERENCES

CHAPTER 1 (page 1)

1. World Bank, *World Development Report 1978* (World Bank: Washington DC, 1978).

2. During preparation of this book a request was made to an international agency for *ex-post* evaluations of irrigation projects. Although they have assisted with planning several hundred irrigation projects over a number of years, they replied: 'Our studies are mainly of a pre-investment nature, rather than post-evaluation of operating irrigation schemes, and we do not therefore have relevant literature based on these projects.' Clearly, they consider pre-investment and *ex-post* evaluation as independent and unrelated activities.

3. Z. Vladisavljevic-Medak, 'Predetermination of investment limits for irrigation schemes' R37, Q33, II. Tenth Congress Proceedings, International Commission for Irrigation and Drainage Athens, 1978.

4. J. C. De Haven, *Journal of American Waterworks Association*, May 1963.

5. J. Hirshleifer and J. W. Milliman, Rand Corporation Publication P3555, p. 2, 1967.

6. R. Rogers and D. V. Smith, 'The integrated use of ground and surface water in irrigation project planning', *American Journal of Agricultural Economics*, **52,** (1), 13–24, 1970.

7. Dew is a relatively unimportant source of crop moisture. Dew and fog harvesting have intrigued several investigators but the unreliability and small quantities obtained in desert areas make it uneconomic; see I. Gindel, 'Irrigation of plants with atmospheric water within the desert', *Nature,* **207,** 1173–5, 1965.

8. I. D. Carruthers, *Irrigation development planning, aspects of Pakistan experience* (University of London, Wye College, 1968).

9. I. D. Carruthers and G. F. Donaldson, 'Estimation of effective risk reduction through irrigation of a perennial crop', *Journal of Agricultural Economics*, **22,** (1), 1971.

10. H. L. Penman, 'Woburn irrigation, 1960–8', *Journal of Agricultural Science*, **75,** (Cambridge, 1970).

11. F. Scholz, 'Migration and nomadism in Baluchistan', *Applied Sciences and Development*, **11,** 90–112 (Tübingen, West Germany, 1978).

12. S. Fresson, 'Public participation on village level irrigation perimeters in the Matam Region of Senegal'. Experience in Rural Development, Occasional Paper 4 (OECD Development Centre: Paris, 1978).

13. R. F. Stoner, private communication.

14. C. J. George, 'The role of the Aswan High Dam in changing the fisheries of the south-eastern Mediterranean', chapter 10 in *The Careless Technology*, M. T. Farvar and J. P. Milton (eds) (The Natural History Press, New York, 1972).

15. From 'A ballad of ecological awareness', printed in Farvar and Milton (eds), *op. cit.*

16. This theme is developed in the paper, R. W. Palmer-Jones and I. D. Carruthers, 'Agricultural water use'. Paper to UNECWA Seminar on *Technology transfer and change in the Arab Middle East*, Beirut, 1977 (Published by Pergamon Press in 1979, edited by A. B. Zahlan).

17. Letitia E. Obeng, 'Too much or too little'; *Ceres*, July–Aug. 1975.

18. O. Aresvik, 'The role of international research centers in the strategy for agricultural development in the less-developed countries', *European Review of Agricultural Economics*, **2–4,** 495–520, 1975.

19. C. Hewitt, *Modernizing Mexican agriculture* (United Nations Research Institute for Social Development, Geneva, 1976).

20. F. Stewart, 'Technology and underdevelopment', *ODI Review,* **1,** 92–105, 1977.

21. V. R. Ruttan, 'Induced innovation and agricultural development', *Food Policy*, **2** (5), 196–216, 1977.

22. A review of the controversy between institutional and economic or technological determinism is contained in V. W. Ruttan 'Induced institutional change', chapter 12, in H. Binswanger and V. W. Ruttan, *Induced Innovation: Technology, Institutions and Development* (Johns Hopkins University Press, Baltimore, 1978).

23. V. N. Rao, 'Linking irrigation with development', *Economic and Political Weekly,* **13,** 24, June 1978.

CHAPTER 2 (page 18)

1. Professor G. E. Blackman, private communication.

2. This section is a very simplified discussion of the scientific understanding of plant–water relations which was first published as an annexe to R. Palmer-Jones and I. Carruthers, 'Agricultural water use'. Seminar on *Technology transfer and change in the Arab Middle East* (UNECWA, Beruit, 1977). (Proceedings published in 1979 by Pergamon Press, edited by A. B. Zahlan.) For a much more comprehensive discussion, see R. O. Slatyer, *Plant–water Relationship* (Academic Press, New York, 1967); or T. L. Hsiao, 'Plant responses to water stress', *Annual Review of Plant Physiology*, 1973.

3. C. Y. Sullivan and J. D. Eastin, 'Plant physiological response to water stress', *Agricultural Meteorology,* **14,** 113–27, 1974 (Elsevier Science Publication Company).

4. An exception occurs on cracking soils where water penetrates to refill the soil around the cracks first.

5. O. T. Denmead and R. H. Shaw, 'Availability of soil water to plants as affected by soil moisture content and meteorological conditions', *Agronomy Journal,* **54,** 385–90, September–October 1962.

6. Professor R. L. Wain, private communication.

7. Sheena Wilkins, Henry Wilkins and R. L. Wain, *Nature,* **259,** (5542), 392–94, February 1976. Report on successful laboratory experiments with powerful growth hormones which encourage seedling root penetration in compacted soils in an article entitled 'Chemical treatment of soil alleviates effects of soil compaction on pea seedling growth'.

8. A. Larqué-Saavedra and R. L. Wain, 'Abscisic acid as a genetic character related to drought tolerance', *Annals of Applied Biology,* **83,** 291–27.

9. E. A. Hurd, 'Plant breeding for drought resistance', chapter 8 in *Water Deficits and Plant Growth.* T. T. Kozlowski (ed.) vol. iv (Academic Press Inc., New York, 1976).

10. *Potential transpiration*, MAFF Technical Bulletin (1) 16 (HMSO, London, 1967).

11. H. L. Penman, 'Natural evaporation from open water, bare soil and grass'. *Proceedings of the Royal Society, London (A),* **193,** 120-45, 1948.

12. C. W. Thornthwaite, 'An approach towards a rational classification of climate, *Geographical Review,* **38,** 55–95 (1948).

13. H. F. Blaney and W. D. Criddle, 'Determining water requirements in irrigated areas from climatological and irrigation data', USDA. Soil Conservation Service, Technical Paper 96 (1950). For a review of the technical aspects of Penman and others, see R. M. Hagen, H. R. Haise and T. W. Edminster (eds), *Irrigation of arid lands*, American Society of Agronomy Monograph 11 (Wisconsin, 1967); and FAO/UNESCO, *Irrigation, drainage and salinity*, chapter 8 (Rome, 1973).

14. See, for a modified Penman method devised for India FAO Irrigation and Drainage Paper 24 (Revised 1977); also 'Estimation of water requirements of crops', FAO/World Bank Cooperative Programme (Rome, 1976).

15. P. Quesnel, *Comptes Rendus de l'Académie d'Agriculture de France* (June 1962).

16. S. L. Rawlins, 'Management of water content and the state of water in soils', chapter 1 in *Water deficits and plant growth*. T. T. Kozlowski (ed.) *Op. cit.*

17. This refers, of course, to water loss per day; crops with a longer growing season naturally require more water in total.

18. A. J. Rutter, 'Water consumption by forcsts' in *Water Deficits and Plant Growth*, T. T. Kozlowski (ed.) vol. ii (1968), *Op. cit.*

19. R. Palmer-Jones, 'Some aspects of the economics of irrigating tea in Malawi', unpublished Ph.D. thesis, University of Reading, 1975.

20. W. R. Gardner, 'The relation of root distribution to water uptake and availability', *Agronomy Journal*, **56,** (1) 41–5, 1964.

21. *Irrigation by sprinklers*, FAO Development Paper 65 (1960). P. H. Nye (Department of Soil Science, University of Oxford) also gives, in a private communication, 60 cm for groundnuts and 240 cm for lucerne (alfalfa). However, the plant obtains decreasing proportions of its water requirements from these deeper roots.

22. E.g. for lettuce, 30 cm × 0·4 mm of moisture/cm = 1·2 cm; for sorghum, 180 cm × 1·75 mm of moisture/cm = 31·5 cm.

23. Sir E. W. Russell, *Soil conditions and plant growth*, 8th ed. (London, 1950).

24. Sir E. W. Russell, private communication.

25. R. K. Schofield, *Translations of the 3rd International Congress of Soil Science,* **21,** 37, 1935.

26. J. Gale and A. Poljakoff-Mayber, 'Plastic films on plants as anti-transpirants', *Science,* **156,** 650, 1967.

27. A. Abou-Khaled, R. M. Hagan and D. C. Davenport, 'Effect of Kaolinite as a reflective anti-transpirant', *Water Resource Research,* **6,** (1), 280, 1970.

28. J. Gale, A. Poljakoff-Mayber, I. Nir and I. Kahane, 'Preliminary trials of the application of anti-transpirants under field conditions to vines and bananas', *Australian Journal of Agricultural Research,* **15,** 929, 1964.

29. R. F. Wynn, 'A note on costing water supply for irrigation', *Sudan Agricultural Journal*, **1** (1), June 1965.

30. H. L. Penman, private communication.

31. Bernard, Association Internationale d'Hydrologie, Publication no. 38.

32. Entwicklungsländer Institut, Berlin, Seminar, August 1967.

33. Government of India, Ministry of Agriculture, *Studies in the Economy of Farm Management* (Madras, 1954–5).

34. G. Mohammad, 'Development of irrigation agriculture in East Pakistan: some basic considerations', *Pakistan Development Review,* **5,** 1–53, 1965.

35. University of Hawaii, Agricultural Experimental Station, *Agricultural Economic Report 72*, 1966.

36. Ruakura Institute, private communication.

37. *Bewasserungs Anlagen* (Zurich, 1944).

38. Communication by State government officials during field visit.

39. *Evaluation of major irrigation projects*, Programme Evaluation, Government of India, 1965.

40. M. F. Ali, 'Performance of the Ganges-Kobadak Project—a case study'. Paper to ODI Workshop on *Choices in Irrigation Management* (University of Kent, September 1976).

41. *Irrigation, drainage and salinity*, 230, FAO/UNESCO, Rome, 1973.

42. S. K. De Datta, K. A. Gomez, R. W. Herdt, and R. Barker, *A Handbook on the Methodology for an Integrated Experiment-survey on Rice Yield Constraints* (The International Rice Research Institute, Los Banos, Laguna, Philippines, 1978).

43. C. V. Moore, 'A general framework for estimating the production function for crops using irrigation water', *Journal of Farm Economics,* **43,** 876–88, 1961.

44. H. P. Mapp, V. B. Eidan, J. F. Store and J. M. Davidson, 'Stimulating soil-water and atmospheric stress-crop yield relationships for economic analysis', Agricultural Experiment Station, Oklahoma State University Technical Bulletin T-140 (February 1975). R. W. Hexem and E. O. Heady, *Water production functions for irrigated agriculture* (Iowa State University Press, Ames, Iowa 1978).

45. S. Reutlinger and J. Seagraves, 'A method of appraising irrigation returns', *Journal of Farm Economics,* **44,** 837–42, August 1962.

46. E. Wisner, 'A modified soil moisture budget for irrigation planning in humid regions', Paper to American Society of Agronomy, November 1961.

47. C. W. Beringer, 'Economic model for determining the production function for water in agriculture', Giannini Foundation Research *Report*. 240 (University of California, Berkeley, 1961).

48. I. D. Carruthers, *Irrigation development planning* (University of London, Wye College, 1968).

49. R. Barker, 'Yield and fertilizer input' in *Changes in Rice Farming in Selected Areas of Asia* (International Rice Research Institute, Los Banos, Philippines, 1978).

50. J. Shalhevet, A. Mantell, N. Bielorai and D. Shimshi, *Irrigation of field crops under semi arid conditions* (International Irrigation Information Center Publication, Bet Dagen, Israel, 1976).

51. R. W. Palmer-Jones, 'Estimating irrigation crop response from data on unirrigated crops', *Journal of Agricultural Economics,* **53** (1), 1976; and R. W. Palmer-Jones, 'Irrigation system operating policies for mature tea in Malawi', *Water Resources Research,* **13** (1), 1–7, 1977.

52. Académie d'Agriculture de France, 960–61, 1969.

53. Doctoral dissertation, University of California (Davis), 1972.

54. *Journal of Australian Institute of Agricultural Science* (November 1957).

55. Winchmore Irrigation Research Station, *Technical Report 5.*

56. New Zealand Water Conference, *Proceedings* (1970).

57. International Grassland Congress, 1970.

58. Hutton, Ruakura Farmers' Conference, June 1973.

59. J. Flinn, *Farm Management Bulletin 1*, 230, University of New England (University of Armidale, NSW).

60. Jhaveri Dairy Farm, Ahmedabad, private communication.

61. I. D. Carruthers, *Irrigation development planning, op. cit.*

62. W. Falcon and C. Gotsch, Harvard Centre for International Affairs, *Economic Development Report 11*, (1969).

63. Ghulam Mohammad, 'Development of irrigated agriculture in East Pakistan: some basic considerations', *Pakistan Development Review,* **6** (3), 315–75, Autumn 1966.

64. C. Gotsch, 'A programming approach to some agricultural policy problems in West Pakistan', *Pakistan Development Review,* **8,** Summer 1968.

65. E. J. Winter, 'Irrigation of vegetables in Britain', *Outlook on Agriculture,* **5,** (4), 1967.

66. 'Elements of irrigation planning', *Annales Agronomiques,* **65,** 1961.

67. D. Yaron, *Economic criteria for water resource development and allocation* (Hebrew University of Jerusalem, 1966).

68. Winchmore Irrigation Research Station, New Zealand, *Technical Report* 13, 1970.

69. H. G. Hogg, L. B. Rankine and J. R. Davidson, 'Estimating the productivity of irrigation water', *USDA Agricultural Economics Research,* **22** (1), 12–7, January 1970.

70. J. Flinn, *op. cit.*

71. D. Denmead and R. Shaw, 'The effects of soil moisture stress at different stages of growth on the development and yield of corn', *Agronomy Journal,* **52,** 5272–4, 1960.

72. E. F. Henzell and B. Stirk, 'Effects of nitrogen and water on the growth of pasture grasses', *Australian Journal of Experimental Agriculture,* **3,** 307–13, 1963.

73. B. S. Minhas, K. S. Parikh, and T. N. Sririvasan, Indian Statistical Institute, New Delhi, *Discussion Paper 59*, and 'Toward the structure of a production function for wheat yields with dated inputs of irrigation water', *Water Resources Research,* **10** (3), 1974.

74. A bibliography on response is presented in Raymond L. Anderson and A. Maass, *A simulation of irrigation systems*. Technical Bulletin 1436, ERS, USDA (revised edition 1974).

75. J. A. Dawson, 'The productivity of water in agriculture', *Journal of Farm Economics* (Proceedings), (1957).

76. Beringer, *op. cit.*

77. C. V. Moore, 'A general analytical framework for estimating production function for crops using irrigation water', *Journal of Farm Economics,* **43,** 876–88, 1961.

78. D. Yaron, G. Strateener and M. Weisbrod, 'Wheat response to soil moisture and the optimal irrigation policy under conditions of unstable rainfall', *water Resources Research,* **9** (5), 1973.

79. Minhas *et al., op. cit.*

80. J. C. Flinn and W. F. Musgrave, 'Development and analysis of input–output relations for irrigation water', *Australian Journal of Agricultural Economics,* **11,** 1–19, 1967; and J. C. Flinn, 'Simulation of crop-irrigation systems' in J. Dent and J. R. Anderson, *Systems Analysis in Agricultural Management* (Wiley, New York, 1971).

81. R. W. Palmer-Jones, 'Some aspects of the economics of irrigating tea in Malawi', unpublished Ph.D. thesis, University of

Reading, 1975. (See also Palmer-Jones, 1976 and 1977, *op. cit.* note 51.)

82. G. W. Irvin, 'Irrigation planning at the farm level using linear programming techniques', Report to Ministry of Overseas Development, 1973, mimeo.

83. US President's Science Advisory Committee, *The world food problem,* **2,** 458, 1967.

84. H. L. Penman, 'Natural evaporation from open water, bare soil and grass', *Proceedings of the Royal Society, London (A),* **193,** 120–45, 1948.

85. A. J. Low and E. R. Armitage, 'Irrigation of grassland', *Journal of Agricultural Science,* **52,** 256–62, 1959.

86. Dr Quershi, Lyallpur Agricultural Research Institute, privately communicated.

87. R. Fernandez and R. J. Laird, 'Yield and protein content of wheat in central Mexico as affected by available soil moisture and nitrogen fertilization', *Agronomy Journal,* **51,** 33, 1959.

88. J. Nix, *Cambridge Farm Economics Branch Report 55*; and *Agriculture*, May 1960.

89. Grassland Research Institute, *Annual Report*; 1957–8 and 1958–9.

90. Entwicklungsländer Institut, Berlin, Seminar, August 1967.

91. World Bank, *World Development Report 1978* (Washington DC, 1978).

92. K. A. Gomez, 'On farm assessment of yield constraints: methodological problems (1977). Pages 1–16 in *Constraints to high yields on Asian rice farms: an interim report* (International Rice Research Institute, Los Banos, Philippines).

93. T. Anden-Lacsina and R. Barker, 'The adoption of modern varieties' in *Changes in rice farming in selected areas of Asia*, IRRI, *Ibid.* 1978.

94. For a comprehensive treatment of response analysis in agriculture, see J. L. Dillon, *The Analysis of Response in Crop and Livestock Production*, 2nd ed. (Pergamon Press, Oxford, 1977).

CHAPTER 3 (page 68)

1. Statistics from R. L. Nace in *Introduction to Geographical Hydrology*, R. J. Chorley (ed.) (Methuen, London 1971).

2. Tixeront, *The Future of Arid Lands* (American Association for the Advancement of Science, 1956).

3. Rural Reconstruction Commission, 1944.

4. H. Roederer, 'Réflections sur les relations précipitations–écoulement', pp. 21–31 of book, quoted by J. M. Houston, *The Significance of Irrigation in Morocco's Economic Development, Geographical Journal,* **120,** 320, September 1954.

5. International Bank for Reconstruction and Development, *The*

Economic Development of Ceylon (Ceylon Government Press, 1952).

6. Wald und Wasser, 1958.

7. Dr Leyton, Forestry Department, Oxford, private communication.

8. For a review of the problems of watershed management as it affects irrigated regions, see CENTO *Watershed Management.* Report on a seminar held at Peshawar, Pakistan (1977).

9. Merendi, *Banco di Roma Review*, 1957.

10. UK Water Resources Board, *Fifth Annual Report* (HMSO, London, 1969).

11. Chang Chih-yi, *Geographical Review*, 1949.

12. J. D. Lang, 'Australian water resources with particular reference to water supplies in Central Australia', *Journal of the Institute of Engineers, Australia*, 31 March 1946.

13. *The Australian Environment* (Commonwealth Scientific and Industrial Research Organization, Melbourne, 1960).

14. P. Lieftinck, A. R. Sadove and T. C. Creyke, *Water and Power-Resources of West Pakistan* (Johns Hopkins University, Baltimore, Ma (for World Bank), 1968).

15. See Arthur Maass, Maynard M. Hufschmidt, Robert Dorfman, Harold A. Thomas Jr, Stephen A. Marglin and Gordon Maskew Fair, *Design of Water Resource Systems* (Cambridge, Ma, Harvard University Press, 1962); Maynard M. Hufschmidt and Myron B. Fiering, *Simulation Techniques for Design of Water Resource Systems* (Cambridge, Ma, Harvard University Press, 1966); Myron B. Fiering, *Streamflow Synthesis* (Cambridge, Ma, Harvard University Press, 1967).

16. For a recent comparative study of the behaviour of American and Spanish irrigators under various levels of water availability, see A. Maass and R. L. Anderson, *And the desert shall rejoice: conflict, growth and justice in arid environment* (MIT Press, Cambridge, Mass, 1978).

17. P. Lieftinck, A. Sadove and T. Creyke, *op. cit.*

18. J. D. Tothill (ed.), *Agriculture in the Sudan*, p. 594 (Oxford University Press, 1948).

19. One author once acted as an advocate at a legal enquiry on behalf of a Hampshire farmer who had been refused permission by the Thames Conservancy to irrigate his grassland. The Conservancy officials were quite candid—they needed all the water in this particular tributary to dilute the increased sewage flow from a newly built-up area. Treating the sewage, they said, would be far too expensive.

Prince Charles, after a swim at a beach near Melbourne, said that he felt that he had been bathing in diluted sewage. This remark caused a considerable stir in local political circles. But it was an accurate description.

20. Department of the Environment, *Hydraulics Research Station, 1974* (HMSO, London, 1975).

21. Pierre Gouro, *Annuaire du Collège de France*, p. 301 (1966).

22. US Government, *Select Committee on National Water Resources* (US Senate, Committee Print 32, 1960).

23. McClellan, *The future of arid lands* (American Association for the Advancement of Science, 1956).

24. Kakuei Tanaka, *Building a New Japan*, p. 150. Estimates by MITI Institute (Simul Press Inc., Tokyo, 1973).

25. Water Resources Board, *Fifth Annual Report* (HMSO, London, 1969).

26. R. Colas, 'Le Problème de l'eau', *Population,* **19,** 31–54 (1964).

27. *Problems of minor irrigation*, Government of India Planning Commission, PEO Publication 40, 138–43 (1961).

28. Ghulam Mohammad, 'Waterlogging and Salinity in the Indus Plain: a critical analysis of some of the major conclusions of the Revelle report', *Pakistan Development Review*, **4,** (3), 357–403, 1964.

29. S. Arlosoroff, *Israel—a model of efficient utilization of a country's water resources*. Israel Thematic Paper 1 to UN Water Conference, Mar del Plata, Argentina, March 1977.

30. D. H. Perkins, *Agricultural Development in China* (Aldine, Chicago, 1969).

31. H. W. Henle, *Report on China's Agriculture*, UN/FAO, p. 137 (1974).

32. D. Twitchett and P. J. Geelan, *The Times Atlas of China* (Quadrangle, 1974).

33. O. L. Danson, *Communist China's agriculture* (Praeger, New York, 1970).

34. K. S. Murthy, 'Interregional water transfers: case study on India', *Water Supply and Management,* **2,** 117–25, 1978.

35. World Bank, *World Development Report 1978* (Washington, DC, 1978).

36. V. Ruttan, *Economic demand for irrigated acreage* (Johns Hopkins University Press, Baltimore, Ma, 1965).

37. Hunting Technical Services and Sir M. MacDonald and Partners, *Lower Indus Project Report,* **2** (for Water and Power Development Authority, Government of Pakistan, 1966).

38. J. C. Day, *Managing the Lower Rio Grande* (University of Chicago, Department of Geographical Research, paper 125, 1970).

39. G. Garbrecht, 'Ways of improving the efficiency of water utilisation in irrigation systems', *Applied Sciences and Development,* **8** (Institute for Scientific Co-operation, Tübingen).

40. For a review of technical material relating to drip irrigation techniques, see *Proceedings Second International Drip Irrigation Congress*, (San Diego, California, July 1974).

41. A. J. Cooper, 'Crop production in recirculating nutrient solutions', *Scientia Horticulturae*, **3,** 251–58, 1975.

42. Burton, *Murray Waters Symposium* (Australian Academy of Science, 1971).

43. J. Ryan, University of Sydney M.Sc. Thesis, 1968.

44. NSW Irrigation Research and Extension Committee, Annual Meeting, 1966.

45. R. L. Tonz, 'Future of irrigation in the humid area', *Journal of Farm Economics,* **40,** 646, 1958.

46. C. Finney, University of Reading Agricultural Economics Department, Development Study 11, p. 107.

47. *Pakistan Development Review* (Autumn 1964, *op. cit.*).

48. International Rice Research Institute Annual Report, p. 142, (1963).

49. Ghulam Mohammad, 'Development of irrigation agriculture in East Pakistan: some basic considerations', *Pakistan Development Review,* **6,** 315–75, 1966.

50. Entwicklungsländer Institut, Berlin, Seminar, August 1967.

51. R. Colas, *op. cit.*

52. M. G. Bos, 'Some influences of project management on irrigation efficiencies' in E. B. Worthington (ed.) *Arid Land Irrigation in Developing Countries* (Pergamon Press, Oxford, 1977).

53. M. G. Bos and C. Storsbergen, 'Irrigation project staffing' R24Q35, ICID *10th Congress Proceedings* (Athens, 1978).

54. M. G. Bos and J. Nugteren, *On Irrigation Efficiencies* (Publication 19, Wageningen: International Centre for Land Reclamation and Improvement, 1974).

55. R. C. Calvert and R. F. Stoner, *Khairpur Tubewell Project—pumping groundwater for irrigation and drainage*, Sir M. MacDonald and Partners, Cambridge, 1974 (mimeo).

56. P. E. Naylor, 'Control of waterlogging and salinity in West Pakistan', *International Journal of Agrarian Affairs,* **4** (1), 1–12, October 1963.

57. R. Revelle, *Report on land and water development in the Indus Plain* p. 411 (US Government 1964).

58. The amount of leaching water required depends upon the salinity of the water percolating down to the water-table. It ranges from 6 per cent total water with 300 parts per million dissolved salts to 20 per cent with 1000 parts per million.

59. Reserve Bank of India, private communication.

60. C. Finney, *Lahore Symposium on Water Resource Development*, 1969; and *Pakistan Development Review*, Winter 1968.

61. Guttridge, Haskins and Davey, *Report to River Murray Commission* (1970).

62. Entwicklungsländer Institut, Berlin, Seminar, August 1967.

63. C. V. Moore, J. H. Snyder and P. Sun, 'Effects of Colorado River water quality and supply on irrigated agriculture', *Journal of Water Resources Research,* **10,** April 1974.

64. K. C. Channabasappa 'Desalination application in Western Asia', in *Technology Transfer and Change in the Arab World*, A. B. Zahlan (ed.) (Pergamon Press, Oxford, 1978).

65. M. Kantor, 'Non-conventional water resources: some advances in their development', supporting document to UN Water Conference, Mar del Plata, Argentina, 1977.

CHAPTER 4 (page 97)

1. A detailed proposal for an induced recharge scheme using riverain wells has been proposed for the Indus River by P. Johnson and R. F. Stoner. Cf. *Proceedings of 1975 Congress of the International Commission for Irrigation and Drainage* (Moscow).

2. Many public agencies expect beneficiaries to pay only operating and maintenance costs. To the degree that O and M costs are true short-run marginal costs, this is an efficient pricing policy. In the case of tubewells, O and M costs represent a high proportion of total costs, and therefore this policy will have attractions to those interested in maximising financial returns.

3. FAO *Medium-term Plan for Development of Ground Water Resources of East Jordan* (Amman, 1974).

4. E. J. Clay, *Choice of techniques: a case study of tubewell irrigation*, based on unpublished D.Phil. thesis, University of Sussex, 1974 (mimeo, 1975).

5. T. V. Moorti and J. W. Mellor, 'A comparative study of costs and benefits of irrigation from state and private tubewells in Uttar Pradesh'. *Indian Journal of Agricultural Economics,* **4,** 181–9, 1973.

6. E. S. Clayton, I. D. Carruthers and F. Hamawi, *Wadi Dhuleil Jordan—an ex-post evaluation*. Occasional Paper 1 (Agrararian Development Unit, Wye College, 1974).

7. For recent documentation of this problem, see M. A. Hamid, S. K. Saha, M. A. Rahman and A. J. Khan, *Irrigation technologies in Bangladesh* (Department of Economics, Rajshahi University, Bangladesh, 1978).

8. Report of the Panel of Economists on the Preliminary Draft of the Two Year Plan, Dacca, March 1978 (quoted by Hamid *et al.*).

9. Credit for the engineering input in this section must go to Roy Stoner and his colleagues at Sir M. MacDonald and Partners. One author (Carruthers) has been a colleague in Pakistan, Bangladesh and Saudi Arabia.

10. In engineering practice, minimum annual cost may be used as a criterion. Minimum present value of capital and recurrent costs is a preferable criterion because it allows the precise timing of operation and maintenance to be taken into account.

11. R. F. Stoner, D. M. Milne and R. J. Lund, 'Economic design of wells in deep aquifers'. *Proceedings of the International Commission on Irrigation and Drainage* (Moscow, 1975).

12. J. W. Thomas, 'The choice of technology for irrigation tubewells in East Pakistan: analysis of a development policy decision' in C. Timmer, J. W. Thomas, L. T. Wells and D. Morawetz, *The choice of technology in developing countries* (Harvard Studies in International Development, Cambridge, Mass., 1975).

13. S. Biggs, C. Edwards and J. Griffith, 'Irrigation in Bangladesh. On contradictions and under-utilized potential' (mimeo paper to Chittagong Economic Association, April 1977); copies available from the Institute for Development Studies, University of Sussex.

14. E. F. Renshaw, 'The management of groundwater reservoirs', *Journal of Farm Economics*, May 285, **45,** 2, 1963.

15. Nazim A. Kazmi, 'Utilization of groundwater potential', *Economic and Political Weekly*, **13,** 17, April 1978.

CHAPTER 5 (page 118)

1. This index was based upon various specified grades of food, livestock, grains, fats and oils, non-food agricultural products, timber, metals and minerals and fertilizers. It was forecast that the next five years would see an increase in international prices as follows:

1974 = 100·0	1976 = 122·4	1978 = 143·4
1975 = 111·5	1977 = 132·8	1979 = 154·2

At the time of writing (1978) these forecasts have proved to be reasonably accurate. Published sources of international price trends are scanty and delayed and there are important differences between countries. To date the most recent UN *Statistical Yearbook* (1976) has price information only for 1975. This shows, for example, that general wholesale prices increased from 1974 to 1975 by 8 per cent in Egypt, 2 per cent in India, 11 per cent in Iraq, 22 per cent in Pakistan, and 8 per cent in USA. Building materials rose slightly faster than the general trend. Farm product prices also rose slightly faster in most countries. To up-date the $74 values in this book to 1979 levels the reader is recommended to apply a factor of 1·5. This will give an indication in current values, in the right direction, and an approximate order of magnitude.

2. *UN Statistical Yearbook*, United National Department of Economic and Social Affairs, New York, published annually.

3. I. B. Kravis, *et al.*, *A system of international comparison of gross production and purchasing power* (Johns Hopkins University, Baltimore, Md, 1975); and S. N. Braithwaite, 'Real income levels in Latin America', *Review of Income and Wealth*, **14,** (2), 113–82, June 1968.

4. J. D. Tothill (ed.), *Agriculture in the Sudan*, p. 608 (Oxford University Press, London, 1948).

5. A. R. Tainsh, private communication.

6. Dias, *Tropical agriculturalist* (Ceylon, 1956).

7. C. Dakshinamurti, A. M. Michael and Shri Mohan, *Water Resources of India and Their Utilization in Agriculture* (Water Technology Centre, IARI, New Delhi, 1973).

8. R. L. Sansom, 'The impact of an insurgent war on the traditional economy of the Mekong River Delta region of South Vietnam', unpublished D.Phil. thesis, Oxford University, 1969.

9. Tothill, *op. cit.*

10. W. D. Hopper, 'Allocation efficiency in a traditional Indian village', *Journal of Farm Economics,* **47** (3), 611–24, August 1965.

11. S. P. Dhondyal, D.Sc. thesis, Delhi University.

12. Shastri, *Economic Weekly*, India, 29 October 1960.

13. Ghosh, *International Statistical Institute Proceedings* (1951).

14. S. S. Wilson, Engineering Laboratory, Oxford, private communication.

15. International Institute of Agriculture, *Monthly Review*, June 1913.

16. Sansom, *op. cit.*

17. S. Biggs, Institute for Development Studies, University of Sussex, private communication.

18. Federazione delle Associazioni Scientifiche e Techniche, Milan.

19. Ghulam Mohammad, 'Private tubewell development and cropping patterns in West Pakistan', *Pakistan Development Review,* **51,** 1965; and also Ghulam Mohammad, 'Waterlogging and salinity in the Indus Plain: A critical analysis of some of the major conclusions of the Revelle report', *Pakistan Development Review,* **4** (3), pp. 357–403, 1964, Karachi.

20. C. E. Finney, *Farm power in West Pakistan* (University of Reading, Agricultural Department, Development Study 11, 1972).

21. Ghulam Mohammad, 'Private tubewell development', *op. cit.*

22. Government of Tunisia, *Estimation des coûts de production et des revenus dans la Tunisie du Nord, 1963.*

23. *Convengno sul probleme delle acque in Italia* (Milan, 1965).

24. A. S. Sirohi and W. H. Pine, 'Irrigation with restraints on land and water resources', *Land Economics,* **45,** May 1969.

25. William E. Martin and Thomas Archer, 'Cost of pumping irrigation water in Arizona 1891 to 1967', *Water Resources Research,* **7,** no. 1, pp. 23–31, February 1971.

26. J. A. Dawson, 'The productivity of water in agriculture', *Journal of Farm Economics*, **29,** (5), 1957, proceedings.

27. P. Vicinelli, 'Irrigation in Italy', *International Journal of Agrarian Affairs,* **3,** 205–32, January 1963.

28. Nalson and Partier, *Irrigation on the Gascoyne River* (University of Western Australia Press).

29. Sansom, *op. cit.*

30. Rowland, *Africa and Irrigation* (Wright Rain Ltd, 1961).

31. S. S. Wilson, Oxford University Department of Engineering Science, private communication.

32. R. F. Wynn, 'A note on costing water supply for irrigation', *Sudan Agricultural Journal,* **1** (1), 1965.

33. I. D. Carruthers, *Irrigation Development Planning: Aspects of Pakistan Experience*, Agrarian Development Studies 2 (Wye College, Ashford, UK, 1968).

34. International Labour Office, 'Appropriate construction technology for water control and irrigation works in developing countries', ILO World Employment Programme contribution to UN World Water Conference, Mar del Plata, Argentina, 1977.

35. J. M. Healey, *The Development of Social Overhead Capital in India, 1950–60* (Blackwell, Oxford, 1965).

36. *Select Committee of National Water Resources*, Committee Print no. 32 (US Senate, 1960).

37. Binnie and Partners reported in *Town and Country Planning* (June 1966).

38. Kanwar Sain (then Director of Irrigation, India), private communication.

39. A. Otten and S. Reutlinger, 'Performance evaluation of eight ongoing irrigation projects', Economics Department Working Paper, no. 40 (World Bank, 1969).

40. A. Farouk, *Irrigation in a Monsoon Land* (University of Dacca, 1968).

41. R. C. Manning, *An economic evaluation of irrigation rehabilitation projects in Mexico*, IBRD Report FC-180 (1971).

42. A. K. Mitra, 'An economic appraisal of farm management and cropping patterns', report prepared by FAO for Government of Iraq as part of *Pilot Project in Soil Reclamation and Irrigated Farming Development in the Greater Mussayib Area; Phase II, Iraq* (Rome, 1975).

43. E. S. Clayton, I. D. Carruthers and F. Hamawi, *Wadi Dhuleil Jordan: an ex-post evaluation*, Occasional Paper 1 (Agrarian Development Unit, Wye College, 1974).

44. Shoaib Sultan Khan, *Daudzai Project: A Case Study* (Pakistan Academy for Rural Development, Peshawar, 1975).

45. A. Ellman and G. Pingle, 'Review of irrigation development in the semi-humid tropics', Workshop on Irrigation Management, Commonwealth Secretariat and Government of India, Hyderabad, October 1978.

46. A. K. Mitra, *op. cit.*

47. Colin Clark, *Economics of Irrigation*, 2nd ed. pp. 66–7 (Pergamon Press, Oxford, 1970).

48. J. Nix, *Farm Management Pocketbook* (6th ed. Wye College, 1974 and 9th ed. Wye College, 1978).

49. Private communication.

50. P. Vicinelli, 'Irrigation in Italy', *International Journal of Agrarian Affairs,* **3,** 205–32, January 1963.

51. *Select Committee of National Water Resources, op. cit.*

52. Fortier, *Use of water in irrigation* (1926).

53. *Econometrica*, October 1957.

54. Rand Corporation publication P.3555, 2, 1967.

55. C. W. Howe and F. P. Linaweaver, 'The impact of price on residential water demand and its relation to system design and price structure', *Water Resources Research*, **3,** (1), 13, 1967.

56. D. G. Miller, M. J. Burley and P. A. Mawer, 'The survey of water supply costs', *Chemistry and Industry*, (21), 23 May 1970.

57. A full account of the issues involved in choosing between thermal and hydro-power is provided by H. G. van der Tak, *The Economic Choice between Hydroelectric and Thermal Power Developments*, World Bank Staff Occasional Paper 1 (Johns Hopkins University, Baltimore, Md, 1969).

58. Sovani and Rath, *Economics of a Multiple Purpose River Dam* (Gokhale Institute Publication).

59. D. Lal and P. Duane, 'A reappraisal of Purna Irrigation Project in Maharashtra, India', IBRD (mimeo, 1971).

60. A detailed water resource example is given in J. Price Gittinger, *Economic analysis of agricultural projects*, p. 147 (Johns Hopkins University, Baltimore, Md, 1972).

61. E. J. Waring, 'Supplementary irrigation of pastures in humid areas', *Review of Marketing and Agricultural Economics*, **27,** (4), 239–54, December 1959.

62. A. J. Randall, 'Irrigation of irrigated and dry land agriculture—profitability and product mix', *Review of Marketing and Agricultural Economics*, **37,** (3), September 1969.

63. *Report on Mareeba-Dimbulah Irrigation Project* (Queensland Irrigation and Water Supply Commission, 1952).

64. A. B. Ritchie, Penshurst, Victoria, private communication.

65. Vicinelli, *op. cit.*

66. A. R. Tainsh, Stockholm, private communication.

67. C. H. Pair, 'Will irrigation pay?', *World Crops,* **14** (4), April 1962.

68. Taylor and Ryde, 20th Lincoln College Grassland Conference, Winchmore Irrigation Research Station Reprint 38 (New Zealand, 1970).

69. Professor Seckler, University of California, private communication.

70. Entwicklungsländer Institut, Berlin, Seminar, August 1967.

71. Seckler, University of California, private communication.

72. Y. Lowe, 'Irrigation and settlement schemes', International Conference of Agricultural Economists, University of Sydney, 1967.

73. Merendi, *Banco di Roma Review*, November 1957.

74. Waring, *op. cit.*

75. Nalson and Partier, *op. cit.*

CHAPTER 6 (page 155)

1. I. D. Carruthers and G. F. Donaldson, 'Estimation of effective risk reduction through irrigation of a perennial crop', *Journal of Agricultural Economics*, **22,** (1), 39–48, 1971.

2. R. L. Anderson, 'The irrigation water rental market: a case study', *Agricultural Economics Research,* **13** (2), April 1961.

3. Ghulam Mohammad, 'Private tubewell development and cropping patterns in West Pakistan', *Pakistan Development Review,* **5,** pp. 1–53, 1965.

4. Institute of Agriculture, Anand, Gujarat, privately communicated.

5. J. A. Dawson, 'The productivity of water in agriculture', *Journal of Farm Economics*, **29,** (5), 1957 proceedings.

6. G Gotsch, 'A programming approach to some agricultural policy problems in West Pakistan', *Pakistan Development Review* (Summer 1968).

7. J. Boussard and M. Petit, 'Problèmes de l'accession à l'irrigation', Institut Nationale de la Recherche Agronomique, SCP, Paris, Le Tholonet, 1966.

8. J. A. Dawson, *op. cit.*

9. University of Hawaii Agricultural Experimental Station, *Agricultural Economic Report 72*, 1966.

10. M. Upton, *Irrigation in Botswana*, Reading University, Department of Agricultural Economics Development Study no. 5.

11. C. V. Moore, J. H. Snyder and P. Sun, 'Effects of Colorado River water quality and supply on irrigated agriculture', *Water Resources Research,* **10,** no. 2, pp. 137–44, April 1974.

12. *Prices of Agricultural Products and Selected Inputs in Europe and North America 1973/74.* UN/ECE/AGRI/13 (United Nations, New York, 1975).

13. Professor Mason Gaffney, University of Milwaukee, private communication.

14. Hopper, *op. cit.*

15. Ghulam Mohammad, 'Private tubewell development', *op. cit.*

16. 'Private tubewell development', *op. cit.*

17. W. Falcon and G. Gotsch, Harvard Center for International Affairs, *Economic Development Report 11*, 1969.

18. T. Jewett for Hunting Technical Services, private communication.

19. H. Wittig, Entwicklungsländer Institut, Berlin Seminar, August 1967.

20. R. L. Anderson and A. Maass, International Commission on Irrigation and Drainage Conference (Phoenix, Arizona, 1968, and Mexico City, 1969).

21. Deepak Lal *et al.*, 'Men or machines: A Philippines case study of labour-capital substitution in road construction', ILO World Employment Programme Working Paper (ILO, Geneva, 1974).

22. R. F. Camacho and A. Bottomley, 'The use of input-output analysis to estimate secondary economic benefits of irrigation schemes', *International Commission on Irrigation and Drainage* R28, Q23, II (10th Congress, Athens, 1978).

23. C. Bell and P. Hazell, 'Measuring the indirect effects of an agricultural investment project on its surrounding region' (mimeo, 1978).

24. H. Bergmann and J. Boussard, *Guide to the Evaluation of Irrigation Projects* (OECD, Paris, 1976).

25. J. C. De Wilde, *Experiences with agricultural development in tropical Africa,* pp. 246–7 (Johns Hopkins, University of Baltimore, Md., 1967).

26. This argument in its extreme form was used by the Government of Queensland before Colin Clark resigned from his position as

their economic adviser and who, in attempting to justify a hopelessly uneconomic irrigation project, set against its annual cost (believe it or not) the expected entire gross product of the farms expected to be established, as if not only the labour and enterprise of the farmers, but also fertilizer, equipment, transport, and so on, were to be had free. One would like to be specifically reassured that such ideas are not still cherished by some Australian irrigation advocates.

CHAPTER 7 (page 184)

1. Ideally, these charges should cover the costs of external diseconomies—for example, the costed loss to fishing should be included if the level of a lake reservoir falls too low to sustain fish life. In other words, in principle, external diseconomies should be internalized.

2. Studies of the theory and practice of public utility pricing are numerous. The earliest reference is Jules Dupuit, 'On the measurement of utility of public works', *Annales des Ponts et Chaussées* ser. 2, vol. 8, 1844 (English translation in International Economic Papers 2, London, 1952.) Theoretical aspects of marginal cost pricing are discussed widely and a classic survey article is Nancy Ruggles, 'Recent developments in the theory of marginal cost pricing', *Review of Economic Studies,* **17,** pp. 107–26, 1949–50. Most studies of application to particular industries have been concerned with electricity; e.g., M. Crew, 'Electricity tariffs' in R. Turvey (ed.), *Public Enterprise* (Penguin Books, London, 1968). For water supply the best article is by J. J. Warford, 'Water requirements: the investment decision in the water supply industry (with an appendix by W. Peters)', *Manchester School,* **34** (1966). There are a few examples of the application of the general economic principles to the particular conditions of less developed countries. One relevant study is Nasim Ansari, *Economics of Irrigation Rates—a Study in Punjab and Uttar Pradesh* (Asia Publishing House, London, 1968). A study relating to domestic water supply in a developing country is I. D. Carruthers, 'A new approach to domestic water rating', *Eastern Africa Economic Review*, **4,** (2), 73–96 (December 1972).

3. R. F. Wynn, 'A note on costing water supplies for irrigation', *Sudan Agricultural Journal*, June 1965.

4. J. Hirshleifer, J. C. De Haven and J. W. Milliman, *Water Supply Economics, Technology and Policy* (Chicago, University of Chicago Press, 1960).

5. J. Sagardoy, 'Water charges for agriculture in some selected countries' (FAO, 1973, mimeo).

6. N. Ansari, 'Some economic aspects of irrigation development in Northern India', *International Journal of Agrarian Affairs*, **4,** (1), October 1963.

7. P. Duane, 'A policy framework for irrigation water charges', World Bank Staff Working Paper 218, 1975.

8. J. P. Naegamvala, 'Development of water resources in India for irrigation and hydro power generation', UN Seminar on Water Resources Administration, New Delhi, 1973.

9. Irrigation Commission, Government of India, *Report of the National Irrigation Commission,* **1,** 430, 1972.

10. Ghulam Mohammad, 'Private tubewell development and cropping patterns in West Pakistan', *Pakistan Development Review,* **5,** 1–53, 1965.

11. C. M. Rafiq, M. A. Mian and R. Brinkman, 'Economics of water use on different classes of saline and alkali land in the semi arid plains of West Pakistan', *Pakistan Development Review,* **8** (1), 23–4, 1968 (Karachi).

12. J. C. De Wilde, *Experiences with agricultural development in tropical Africa* (Johns Hopkins University, Baltimore, Md, 1967).

13. Government of Kenya, *National Irrigation Board Annual Report and Accounts 1975–76* (National Irrigation Board, Nairobi, 1977).

14. *Problems of minor irrigation*, Planning Commission, PEO Publication 40, 138–43, 1961.

15. Professor Vanzetti, private communication.

16. J. Rutherford, 'Integration of irrigation and dryland farming in the Southern Murray Basin', *Review of Marketing and Agricultural Economics*, **27,** (3), September 1959.

17. P. Vicinelli, 'Irrigation in Italy', *International Journal of Agrarian Affairs,* **3,** January 1963.

18. Entwicklungsländer Institut, Berlin, Seminar, August 1967.

19. Sagardoy, *op. cit.*

20. A. Ward, 'Economics of irrigation water in the Litani River Basin of Lebanon', *International Journal of Agrarian Affairs*, June 1959.

21. W. Beckett, *Royal Central Asian Society Proceedings*, 1963.

22. Gokhale Institute, Poona, private communication.

23. *Times of India* (Ahmedabad, 13 September 1968).

24. E. F. Renshaw, 'Value of an acre foot of water', *Journal of the American Waterworks Association,* **50,** 303–8, March 1958.

25. C. W. Howe and F. P. Linaweaver, 'The impact of price upon residential water demand', *Water Resources Research,* **3** (1), 1967.

26. McClellan, *The Future of Arid Lands* (American Association for the Advancement of Science, 1956).

27. J. C. De Haven, 'Water supply economics technology and policy', *Journal of the American Waterworks Association,* **55,** 539–47, May 1963.

28. A. Maass and R. L. Anderson, *And the Desert Shall Rejoice: Conflict, Growth, and Justice in Arid Environments* (MIT, 1978).

29. D. C. Taylor, 'Formulating financial policies in large-scale irrigation projects', Land Tenure Center *Newsletter 54*, October–December 1976.

30. J. E. Nickum, Center for Chinese Studies, University of California, Berkeley, private communication.

CHAPTER 8 (page 201)

1. T. Yates, *Canal linings and canal seepage*, Report OD/1 (Hydraulics Research Station, Wallingford, 1975).
2. K. A. Wittfogel, *Oriental despotism* (New Haven, Conn., 1957).
3. R. Chambers and J. Moris (eds), *Mwea—an Irrigated Rice Scheme in Kenya* (Welforum, Munich, 1973).
4. I. D. Carruthers and G. F. Donaldson, 'Estimation of effective risk reduction through irrigation of a perennial crop', *Journal of Agricultural Economics,* **22,** no. 1, January 1971.
5. Glenn Johnson reports similar problems in a sewage disposal project at Michigan State University. 'All that was needed, we thought, was some simple economic theory, appropriate experimental designs, and the application of well-known econometric techniques. However, it soon became painfully clear that the concepts of soil chemistry, soil physics, and soil biology were inadequate. Further, measurement techniques were so lacking that the supply of nitrogen in the soil could not be measured with any degree of accuracy. Also, our economic theories as well as the experimental designs offered by the statisticians and bio-metricians were woefully deficient'. G. L. Johnson, 'The quest for relevance in agricultural economics', *American Journal of Agricultural Economics,* **53** (5), 1971.
6. D. J. Bradley, 'The health implications of irrigation schemes and man-made lakes in tropical environments' in R. Feachem, M. McGarry and D. Mara (eds), *Water, Wastes and Health in Hot Climates* (Wiley, London, 1977).
7. For a complete review of relevant health issues, see N. F. Stanley and M. P. Alpers (eds), *Man-made Lakes and Health* (Academic Press, London, 1975).
8. Government of Kenya, National Irrigation Board *Annual Report and Accounts 1975–6* (NIB, Nairobi, Kenya, 1977).
9. I. Carruthers, 'Contentious issues in planning irrigation schemes', *Water Supply and Management*, vol. 2, pp. 301–8 (Pergamon Press, Oxford, 1978).
10. E.g. I. M. D. Little and J. Mirrlees, *Project Appraisal and Planning for Developing Countries* (Heinemann, London, 1974); UNIDO; *Guidelines for Project Evaluation* (United Nations, New York, 1972); L. Squire and H. G. van der Tak, *Economic Analysis of Projects* (Johns Hopkins University, Baltimore, Md, 1975).
11. For a detailed critique of the more elaborate procedures of cost-benefit analysis, see I. Carruthers 'Applied project appraisal', *ODI Review* **2** (Overseas Development Institute, London, 1977) and I. D. Carruthers and E. S. Clayton, 'ex-post evaluation of agricultural projects—implications for planning' in *Journal of Agricultural Economics*, **22,** (3) 1977. For a review of the theory and techniques for multiple objective planning specifically for the water sector, see D. C. Major, *Multi-objective Water Resource Planning*, American Geophysical Union, Water Resources Monograph 4, 1977.

12. Robert Chambers, 'Simple is practical; a commentary on project appraisal', *World Development,* **6** (2) (Pergamon Press, Oxford, 1978).

13. J. Price Gittinger, *Economic Analysis of Agricultural Projects* (Johns Hopkins Univerity, Baltimore, Md, 1972).

14. L. Poulequin, 'Risk analysis in project appraisal', World Bank Occasional Paper 11, 1971.

15. Little and Mirrlees, *op. cit.*, for details.

16. Cf. J. Price Gittinger, *Economic analysis of agricultural projects* (Johns Hopkins University, Baltimore, Md, 1972) or Little and Mirrless, *op. cit.*, for details of procedure.

17. Ian Carruthers and Neil Mountstephens, 'Integration of socio-economic and engineering perspectives in irrigation design', II International Commission for Irrigation and Drainage, 10th Congress Proceedings R29Q33 (Athens, 1978).

18. N. D. Hameed (ed), *Rice revolution in Sri Lanka* (United Nations Research Institute for Social Development, Geneva, 1977).

19. World Bank, *Development Report 1978, op. cit.*

20. V. R. Ruttan, 'Induced innovation and agricultural development', *Food Policy,* **2** (3), pp. 196–216, 1977.

21. A review of World Bank practice is contained in Warren C. Baum, 'The World Bank project cycle', *Finance and Development,* **15,** 4 December 1978.

22. There are methods by which some socio-cultural constraints can be incorporated into the LP format. A brilliant representation of an economist's view of a sociologist's specification is presented by Leonard Joy, 'An economic homologue of Barth's presentation of economic spheres n Dafur' in *Themes in Economic Anthropology*, R. Firth (ed.) ASA Monograph 6 (Tavistock, London, 1967).

23. For an introduction to LP, see E. O. Heady and W. V. Chandler, *Linear Programming Methods* (Ames: Iowa University Press, 1958). For farm applications, see C. S. Barnard and J. S. Nix, *Farm Planning and Control* (Cambridge University Press, 1973).

24. The amount of leaching water required depends upon the salinity of the water percolating down to the water-table. It ranges from 6 per cent total water with 300 parts per million dissolved salts to 20 per cent with 1000 parts per million.

25. Cf. Carruthers and Donaldson, 1971, *op. cit.*

26. S. Reutlinger, *Techniques for Project Appraisal under Uncertainty* (Johns Hopkins University, Baltimore, Md, 1970); and Ministry of Overseas Development, 'A guide to the economic appraisal of projects in developing countries', chapter 7 (HMSO, London, 1977).

27. Robert Wade, 'Administration and the distribution of irrigation benefits', *Economic and Political Weekly,* **X,** 44 and 45, 1743-7, November 1975; Robert Wade, 'Performance of irrigation projects', *Economic and Political Weekly,* **XI** (3), 63–6, January 1976a; Robert Wade, 'How not to redistribute with growth: the case of India's command area development programme', *Pacific Viewpoint,* **17** (2),

96–104, September 1976b; A. Bottrall, 'Reports of the ODI network on the management of irrigation schemes' (Overseas Development Institute, London, 1975ff).

28. A. Bottrall, 'The management and operation of irrigation schemes in less developed countries', *Water Supply and Management,* **2,** (4), 1978.

29. R. Chambers, 'Identifying research priorities in water development', *Water Supply and Management*, **2,** (4), 1978.

30. R. C. Terry, 'Management information system development project for Bangladesh Agricultural Development Corporation', *Ford Foundation* (Dacca, 1977).

31. Summary of conclusions of workshop on Choices in Irrigation Management organized by Overseas Development Institute at University of Kent, Canterbury, September 1976, and published as 'The World Water Conference. A suggested action programme on irrigation management', *ODI Review,* **1,** 1977, and as part of British Government's submission to World Water Conference, Mar del Plata, Argentina, 1977.

CHAPTER 9 (page 233)

1. R. W. Herdt and J. W. Mellor, 'The contrasting response of rice to nitrogen: India and the United States', *American Journal of Farm Economics,* **45,** February 1964.

2. G. R. Allen, 'Confusion on fertilizers and the world food situation', *European Chemical News*, Large Plants Supplement, October 1974.

3. R. Barker, 'Yield and Fertilizer input', *Interpretive Analysis of Selected Papers from Changes in Rice Farming in Selected Areas of Asia* (International Rice Research Institute, Los Banos, Philippines, 1978).

4. Michael Lipton, 'The technology, the system and the poor: the case of the new cereal varieties'. Paper for the Institute of Social Studies 25th Anniversary Conference, The Hague, 1977.

BIBLIOGRAPHY

GENERAL/REGIONAL

BISWAS, A. K., *United Nations Water Conference: Summary and Main documents* (Pergamon Press, Oxford, 1977).

BRADLEY, D. J., 'The health implications of irrigation schemes and man-made lakes in tropical environments' in R. Feachem, M. McGarry and D. Mara (eds), *Water, wastes and health in hot climates* (Wiley, London, 1977).

DASGUPTA, BIPLAP, *Agrarian Change and the New Technology in India*, United Nations Research Institute for Social Development, Report 77, 2, Geneva, 1977.

DOWNING, T. E. AND GIBSON, M. (eds), *Irrigation's Impact on Society* (University of Arazona Press, Tuscon, 1974).

ELLMAN, A. AND PINGLE, G., 'Review of irrigation development in the semi-humid tropics' (Workshop of irrigation management, Commonwealth Secretariat and Government of India, Hyderabad, October 1978).

Environmental, Health and Human Ecologic Considerations in Economic Development Projects (World Bank, Washington, DC, May 1974).

FARMER, B. H. (ed.), *Green revolution? Technology and change in rice growing areas of Tamilnadu and Sri Lanka* (Macmillan, London, 1977).

FAROUK, A., *Irrigation in a Monsoon Land* (University of Dacca, 1968).

FARVAR, M. T. AND MILTON, J. P., *The Careless Technology: Ecology and International Development* (The Natural History Press, New York, 1972).

FEACHEM, R. G., MCGARRY, M. G. AND MARA, D. D., *Water, Waste and Health in Hot Climates* (Wiley International, London, January 1977).

FUKUDA, H., *Irrigation in the World—Comparative Developments* (University of Tokyo Press, Tokyo, Japan, 1976), 300 p.

FAO, *Labour-saving Ideas, Series 2: Food, Water, Transport*, FAO Economic and Social Development Series 5/4, 1977 (FAO, Rome, Italy, 1977).

GOLDMAN, C. R. *et al.* (eds), *Environmental Quality and Water Development* (W. H. Freeman, San Francisco, 1973).

JAMES, L. D., *Man and Water*, The Social Sciences in Management of Water Resources (University Press of Kentucky, 1974).

MAASS, A. AND ANDERSON, R. L., *And the Desert shall Rejoice: Conflict, Growth and Justice in Arid Environments* (Massachusetts Institute of Technology, 1977).

NIX, J., *Farm Management Pocketbook* (6th edn., Wye College, 1974, and 9th edn., Wye College, 1978).

PACEY, ARNOLD, 'Technology is not enough: the provision and maintenance of appropriate water supplies', *Water Supply and Management (AQUA)*, **1,** (1–2) (Pergamon Press, 1977).

PEREIRA, H. C., *Land Use and Water Resources* (Cambridge University Press, 1973).

SMITH, N., *Man and Water: a history of hydro-technology* (Peter Davies, London, 1976).

STANLEY, N. F. AND ALPERS, M. P., *Man-made Lakes and Human Health* (Academic Press, 1975).

UNITED NATIONS, *Integrated River Basin Development* (Department of Economic and Social Affairs, New York, 1970).

—— ——, *National Systems of Water Administration, ST/ESA/17* (Department of Economic and Social Affairs, New York, 1974).

—— ——, *Management of International Water Resources—Institutional and Legal Aspects, ST/ESA/5* (Department of Economic and Social Affairs, New York, 1975).

—— —— Water Conference Secretariat, 'Resources and needs: assessment of the world water situation', *Water Supply and Management,* **1,** (3) (Pergamon Press, 1977).

—— ——, *Water Development and Management: Proceedings of United Nations Water Conference*, 4-part (Pergamon Press, Oxford, 1977).

VLACHOS, EVAN, *Transfer of Water Resources Knowledge* (Water Resources Publications, Fort Collins, Colorado, 1975).

VOHRA, B. B., *Land and Water Management Problems in India,* (Training division, Department of Personnal and Administrative Reforms, New Delhi, India, 1975).

WIDSTRAND, CARL (ed.), *Water and Society: Conflicts in Development*, Part 1. The social and ecological effects of water development in developing countries (Scandinavian Institute of African Studies, Uppsala, Sweden, 1978, and Pergamon Press, Oxford).

WITTFOGEL, K. A., *Oriental Despotism* (Yale University Press, New Haven, Conn, 1957).

WORLD BANK, *World Development Report 1978* (World Bank, Washington, DC, 1978).

WORTHINGTON, E. B., *Arid Land Irrigation in Developing Countries Environmental Problems and Effects* (Pergamon Press, Oxford, 1977), 492 p.

ZAHLAN, A. B. (ed.), *Technology Transfer and Change in the Arab Middle East* (Pergamon Press, Oxford, 1978).

TECHNOLOGY

Ackermann, W. C. and White, G. F., *Man-made Lakes: Their Problems and Environmental Effects* (American Geophysical Union, Washington, 1973).

Arlosoroff, S., *Irrigation Equipment and Methods: Trends and Forecasts,* Israel Water Commission, Tel Aviv, and FAO, Rome (Italy, 1976).

Ayres, R. S. and Westcot, D., *Water Quality for Agriculture*, Irrigation and Drainage Paper 29 (FAO, Rome, Italy, 1975).

Barreto, G. B., *Irrigation: Principles, Methods and Experimental Procedures*, 2nd ed (Irrigacao: Prubcupios, Metodos e Practica) Campinas, Brazil (Instituto Campineira de Ensino Agricola, 1974) (Portuguese).

Bauzil, A., *Handbook on Irrigation*, 2 vols (Traite d'Irrigation) Ecole Nationale de Genie Rual des Eaux et des Forets (Paris, France, 1975) (French).

Booher, L. J., *Surface Irrigation*, FAO Agricultural Development Paper no. 95 (FAO Land and Water Development Series no. 3, 1974) (FAO, Rome, Italy, 1976), 2nd printing.

Bos, M. G. and Nugteren, J., *Irrigation Efficiency in Small-farm Areas* (International Commission on Irrigation and Drainage, New Delhi, India, 1974).

——, ——, *On Irrigation Efficiencies* (International Institute for Land Reclamation and Improvement, Wageningen, The Netherlands, 1974).

Bresler, E., *Trickle-drip Irrigation: Principles and Application to Water Management in Tropical Soils* (Cornell University, Department of Agronomy, New York, 1975).

Brown, R. H., Konoplyantsev, A. A., Ineson, J. and Kovalevsky, V. S., *Ground-water Studies—an International Guide for Research and Practice* (Unesco Press, Paris, 1975).

Carr, D. P. and Underhill, H. W., *Simulation Methods in Water Development*, Irrigation and Drainage Paper 23 (FAO, 1974).

Dhawan, B. D., 'Economics of groundwater utilization: traditional versus modern techniques', *Economic and Political Weekly,* **10,** (25–6), pp. A32–42, 1975.

Dunne, Thomas and Luna, B. Leopold, *Water in Environmental Planning* (W. H. Freeman, San Francisco, 1978).

FAO. *Small Hydraulic Structures*, Irrigation and Drainage Paper (26, 1–2) (Water Resources, Development and Management Services, Rome, 1975).

Garbrecht, G., 'Ways of improving the efficiency of water utilization in irrigation systems', *Applied Sciences and Development*, **8** (Institute for Scientific Cooperation, Tübingen).

Goldberg, D., Gornat, B. and Rimon, D., *Drip irrigation: Principles, Design and Agricultural Practices*, Drip Irrigation Scientific Publication (Kfar Shmaryahu, Israel, 1976).

HAGAN, R. M., HOUSTON, C. E. AND ALLISON, S. V., *Successful Irrigation: Planning, Development, Management*, 2nd ed. (FAO, Rome, Italy, 1975) (English, French, Spanish).

HUFSCHMIDT, M. MAYNARD AND FIERING, B. MYRON, *Simulation Techniques for Design of Water Resource Systems* (Harvard University Press, Cambridge, Ma, 1966).

ISRAELSEN, D. W. AND HANSEN, V. E., *Irrigation Practice and Principles* (J. Wiley & Sons, Utah State University, 1976).

JENSEN, M. E., *Scientific Irrigation Scheduling for Salinity Control of Irrigation Return Flows* (United States Environmental Protection Agency, 1975).

KARMELI, D. AND KELLER, J., *Trickle Irrigation Design* (Rain Bird, Glendora, Ca., USA, 1975).

KING, L. G. AND HANKS, R. J., *Irrigation Management for Control of Quality of Irrigation Return Flow* (Environmental Protection Agency, Washington, DC, 1973).

KRAATZ, D. B., *Irrigation Canal Lining*, Land and Water Development Series no. 1 (FAO, Italy, 1977).

LEIFTINCK, P., SADOVE, A. R. AND CREYKE, T. C., *Water and Powerresources of West Pakistan* (Johns Hopkins University Press, Baltimore, Ma, 1968) (for World Bank).

MICHAEL, A. M. *et al., Design and evaluation of irrigation methods*, (Water Technology Centre, Indian Agricultural Research Institute, New Delhi, 1972).

——, *Irrigation: Theory and Practice* (Vikas Publishing House, 5 Ansari Road, New Delhi 110002, India).

MINISTRY OF AGRICULTURE, *Irrigation*, 4th ed. (HMSO, London, 1974).

MOORTI, T. V. AND MELLOR, J. W., 'A comparative study of costs and benefits from state and private tubewells in Uttar Pradesh', *Indian Journal of Agricultural Economics,* **28,** 4, 1973.

MURTHY, K. S., 'Interregional water transfers: case study on India', *Water Supply and Management*, **2,** 117–25, 1978.

NACE, R. L., in *Introduction to Geographical Hydrology*, CHORLEY, R. J. (ed.) (Methuen, London, 1971).

NORTON, R. D., BASSOCO, L. M. AND SILOS, J. S., *Appraisal of Irrigation Projects and Related Policies and Investments* (World Bank, Washington, DC, 1974).

PAIR, C. H., HINZ, W. W., REID, C. AND FROST, K. R. (eds), *Sprinkler Irrigation* 4th ed. (Sprinkler Irrigation Association, Silver Spring, Md., USA, 1975).

ROGERS, P. AND SMITH, D. V., 'The integrated use of ground and surface water in irrigation project planning', *American Journal of Agricultural Economics*, **52,** (1), 1970.

STONER, R. F., MILNE, D. M. AND LUND, R. J., 'Economic design of wells in deep aquifers', *Proceedings of the International Commission on Irrigation and Drainage* (Moscow, 1975).

THOMAS, J. W., 'The choice of technology for irrigation tubewells in East Pakistan: an analysis of a development policy desision' in *The*

Choice of Technology in Developing Countries, C. Timmer, J. W. Thomas, L. T. Wells and D. Morawetz (Harvard Studies in International Development, Cambridge, Mass, 1975).

WITHERS, B. AND VIPOND, S., *Irrigation Design and Practice* (B. T. Batsford Press, 1974).

UNITED NATIONS, *Groundwater Storage and Artificial Recharge, Natural Resources/Water Series 2, ST/ESA/13* (Department of Economic and Social Affairs, New York, 1975).

YOUNG, R. A. AND BREDEHOEFT, J. D., 'Digital computer simulation for solving management problems of conjunctive groundwater and surface water systems', *Water Resources Research*, **8,** (3), June 1972.

CROP PRODUCTION AND RESPONSE

ANON, 'Effect of irrigation and soil moisture on cauliflower, leek and onion' (Centre for Agricultural Publishing and Documentation, Wageningen, The Netherlands, 1975) 45 citations.

ASOPA, V. N., GUISE, J. W. B. AND SWANSON, E. R., 'Evaluation of returns from irrigation of corn in a sub-humid climate', *Agricultural Meteorology,* **ii,** 1973.

COOPER, A. J., 'Crop production in recirculating nutrient solutions', *Scientia Horticulture,* **3,** 251–58, 1975.

DENMEAD, D. T. AND SHAW, R. H., 'Availability of soil water to plants as affected by soil moisture content and meteorological conditions', *Agronomy Journal of Science*, 385–90, September–October 1962.

DILLON, J. L., *The Analysis of Response in Crop and Livestock Production*, 2nd ed. (Pergamon Press, Oxford, 1977).

FISCHER, R. A. AND HAGEN, R. M., 'Plant water relations. Irrigation management and crop yields', *Experimental Agriculture*, **1,** (3), 161–77, July 1965.

GINDEL, I., 'Irrigation of plants with atmospheric water within the desert', *Nature,* **207,** 1173–5, 1965.

GOMEZ, K. A., 'On farm assessment of yield constraints: methodological problems'. Pages 1–16 in *International Rice Research Institute Constraints to High Yields on Asian Rice Farms: an Interim Report* (Los Banos, Philippines, 1977).

HAGEN, R. M., HAISE, H. R. AND EDMINSTER, T. W. (eds), 'Irrigation of arid lands', American Society of Agronomy Monograph no. 11 (Wisconsin, 1967); and FAO/UNESCO, *Irrigation, Drainage and Salinity* (Rome, 1973). (Useful for Penman method.) For a modified Penman method devised for India, see FAO 'Irrigation and drainage, paper no. 24' (revised 1977); also 'Estimation of water requirements of crops' (FAO/World Bank Cooperative Programme, Rome, 1976).

HILLEL, D., *Computer Simulation of Soil-Water Dynamics: a Compen-*

dium of Recent Work (International Development Research Centre, Ottawa, Canada, 1977).

HSIAO, T. L., 'Plant responses to water stress', *Annual Review of Plant Physiology*, 1973.

KOZLOWSKI, T. T. (ed.), *Water Deficits and Plant Growth*, 4 vols (Academic Press, New York, 1976).

KUNG, P., *Irrigation Agronomy in Monsoon Asia* (FAO Plant Production and Protection Division, Rome, 1971).

LANGE, O. L., KAPPEN, L. AND SCHULSE, E. D. (eds), *Water and Plant Life: Problems and Modern Approaches* (German Federal Republic, Springer Verlag, Berlin, 1976).

MALTHOTRA, S. *et al.*, 'Impact of irrigation on land utilisation and cropping patterns in a desert region', *Annals of Arid Zone,* **10,** (2–3), 203–14, 1971.

MAPP, H. P. *et al.*, *Simulating Soil Water and Atmospheric Stress–Crop Yield Relationships for Economic Analysis*, Technical Bulletin T-140 (Agricultural Experimental Station, Oklahoma State University, 1975).

MEIDNER, H. AND SERIFF, D. W., *Water and Plants* (Blackie, Glasgow, 1976).

MINISTRY OF AGRICULTURE, *Potential Transpiration*, Technical Bulletin 16 (HMSO, London, 1967).

NAS, *More water for arid lands: promising technologies and research opportunities* (National Academy of Sciences, Washington, DC, USA, 1974).

PARVIN, D. W., 'Estimation of irrigation response from time-series data on non-irrigated crops', *American Journal of Agricultural Economics*, **55,** (1), 73–7, 1973.

SAHOO, B. C. AND PATRO, G. K., 'Studies on the comparative efficiency of selected crop patterns in irrigated high lands of Bhubaneswar', *Mysore Journal of Agricultural Science*, **8(b),** 376–83, 1974.

SALTER, R. J. AND GOODE, J. E., 'Crop response to water at different stages of growth', *Commonwealth Agricultural Bureau of Research Review 2* (Farnham Royal, Bucks., 1967).

SHALHEVET, J. AND KEMBUROV, J., *Irrigation and Salinity: a Worldwide Survey* (International Commission on Irrigation and Drainage, New Delhi, India, 1976), 106 p.

——, MANTELL, A., BIELORAI, H. AND SHIMSHI, D., *Irrigation of Field and Orchard Crops under Semi-arid Conditions* (International Irrigation Information Centre, Bet Dagan, Israel, 1976).

SLATYER, R. O., *Plant-water relationships* (Academic Press, London and New York, 1967).

SULLIVAN, C. Y. AND EASTIN, J. D., 'Plant physiology response to water stress', *Agricultural Meteorology,* **14,** 113–27, 1975 (Elsevier Science Publishing Company).

WINTER, E. J., *Water, Soil and the Plant* (Science in Horticulture Series) (Macmillan, London, 1974).

WISNER, E., 'A modified soil moisture budget for irrigation planning in humid regions'. Paper to American Society of Agronomy, November 1961.

YARON, D., STRATEENER, G., SHIMSHI, D. AND WEISBROD, W., 'Wheat response to soil moisture and the optimal irrigation policy under conditions of unstable rainfall', *Water Resources Research,* **9,** 1145–54, 1973.

ECONOMICS

ABATE, E., *Optimal Allocation and Management of Agricultural Water* (Colorado State University, Fort Collins, Co, USA, 1975).

ANDERSON, R. L., 'The irrigation water rental market: a case study', *Agricultural Economics Research*, **13,** (2), 1961.

ANSARI, N., *Economics of Irrigation Rates. A study on Punjab and Uttar Pradesh* (Asia Publishing House, London, 1968).

BARON, L. I. Z., *Water Supply Constraint: an Evaluation of Irrigation Projects and their Role in the Development of Afghanistan* (International Development Research Centre, Montreal, Canada, 1975).

BERGMANN, H. AND BOUSSARD, J. M., *Guide to the Economic Evaluation of Irrigation Projects* (revised version), (Organization for Economic Co-operation and Development, Paris, 1976), 257 p. (English). Also published in French under the title *Guide de l'evaluation economique des projects d'irrigation.*

BERINGER, C. A., 'Economic model for determining the production function for water in agriculture', Giannini Foundation Research Report 240 (University of California, Berkeley, 1961).

BIGGS, S., EDWARDS, C. AND GRIFFITH, J., 'Irrigation in Bangladesh. On contradictions and under-utilized potential' (mimeo), Paper to Chittagong Economic Association, copies available from Institute for Development Studies, University of Sussex (April 1977).

BOTTRALL, A., 'The management and operation of irrigation schemes in less developed countries', *Water Supply and Management*, **2,** (4), 1978.

CARRUTHERS, I. D., 'Applied project appraisal', *ODI Review,* **2** (Overseas Development Institute, London, 1977).

—— AND CLAYTON, E. S., 'Ex-post evaluation of agricultural projects—implications for planning', *Journal of Agricultural Economics,* **28,** (3), 1977.

—— AND DONALDSON, G. F., 'Estimation of effective risk reduction through irrigation of a perennial crop', *Journal of Agricultural Economics*, **22,** (1), 1971.

—— AND MOUNTSTEPHENS, N., 'Integration of socio-economic and engineering perspectives in irrigation design', International Commission for Irrigation and Drainage, *10th Congress Proceedings*, (Athens, R29 Q33, 1978).

CHAMBERS, R., 'Simple is practical; a commentary on project appraisal',

World Development, **6,** no. 2, 1978 (Pergamon Press, Oxford).

—— AND MORIS, J. (eds), *Mwea: an Irrigated Rice Settlement in Kenya* (German Federal Republic, Weltforum Verlag, Munich, 1973).

CLARK, C., 'The value of irrigation water', *Economic Analysis and Policy*, **2,** (2), 14–18, 1971.

CLAY, E. J., *Choice of Techniques: a Case Study of Tubewell Irrigation* (mimeo, 1975, based on unpublished D.Phil, thesis, University of Sussex, 1974).

——, *Planners' Preference and Local Innovations in Tubewell Irrigation in North-East India*, IDS Discussion Paper 40 (University of Sussex, January 1974).

CLAYTON, E. S., CARRUTHERS, I. D. AND HAMAWI, F., *Wadi Dhuleil Jordan—an Ex-post Evaluation*, Occasional Paper 1 (Agrarian Development Unit, Wye College, 1974).

CUMMINGS, R. G., 'Optimum exploitation of groundwater reserves with saltwater intrusion', *Water Resources Research*, **7,** (6), December 1971.

——, 'Water resource management problems in Northern Mexico', unpublished paper presented at the Workshop on Problems of Agricultural Development in Latin America (Caracas, Venezuela, 17–19 May 1971).

DE DATTA, S. K., GOMEZ, K. A., HERDT, R. W. AND BARKER, R., *A Handbook on the Methodology for an Integrated Experiment Survey on Rice Yield Constraints* (The International Rice Research Institute, Los Banos, Laguna, Philippines, 1978).

DUANE, P., *A Policy Framework for Irrigation Water Charges*, vol. 7 (incl. bibliography, 4 annexes), mimeo (World Bank, Washington, DC, July 1975), 44 p.

DUDLEY, N. J., 'A simulation and dynamic programming approach to irrigation decision making in a variable environment', Agricultural Economics Business Management Bulletin 9 (University of New England, Armidale, NSW, 1970).

——, 'Irrigation planning 4: Optimal inter-seasonal water allocation', *Water Resources Research*, **8** (3), 586–94, 1972.

FLINN, J. C., 'Simulation of crop-irrigation systems', in J. Dent and J. R. Anderson, *Systems Analysis in Agricultural Management* (Wiley, New York, 1971).

—— AND MUSGRAVE, W. F., 'Development and analysis of input-output relations for irrigation water', *Australian Journal of Agricultural Economics*, **11,** 1967.

FRESSON, S., 'Public participation on village level irrigation perimeters in the Matam Region of Senegal', Experiences in Rural Development, Occasional Paper 4 (OECD Development Centre, Paris, 1978).

GARDNER, B. D. AND FULLERTON, H. H., 'Transfer restrictions and the misallocation of irrigation water', *American Journal of Agricultural Economics*, **50,** (3), 1978.

GISSER, M., 'Linear programming models for estimating the agricultural demand function for imported water in the Pecos River Basin', *Water Resources Research*, **6,** (4), August 1970.

GITTINGER, J. PRICE, *Economic analysis of agricultural projects* (Johns Hopkins University Press, Baltimore, Md, 1972).

GOTSCH, C., 'A programming approach to some agricultural policy problems in West Pakistan', *Pakistan Development Review,* **8,** Summer 1968.

GREEN, D. E., *Land of the underground rain: irrigation of the Texas high plains. 1910–70* (University of Texas Press, Austin, USA, 1973).

GYSI, M. AND LOUCKS, D. P., 'Some long run effects of water pricing policies', *Water Resources Research,* **7(b),** 1371–82, 1971.

HADARI, A. M., 'Irrigated agriculture in the Sudan: new approaches to organisation and management', *Indian Journal of Agricultural Economics*, **27,** (2), 23–37, 1972.

HAMEED, N. D. ABDUL, AMERASINGHE, N., PANDITHARATNA, B. L., GUNASEKERA, G. D. A., SELVADURAI, J. AND SELVANAYAHAM, S., *Rice Revolution in Sri Lanka* (United Nations Research Institute for Social Development, Geneva, 1977).

HAMID, M. A., SAHA, S. K., RAHMAN, M. A. AND KHAN, A. J., *Irrigation Technologies in Bangladesh: a Study in some Selected Areas* (Department of Economics, Rajshahi University, Rajshahi, Bangladesh, 1978).

HARTMAN, L. M. AND ANDERSON, R. L., 'Estimating the value of irrigation water from farm sales data in north-eastern Colarado', *Journal of Farm Economics*, **44,** (1), 1962.

HAZELWOOD, A. AND LIVINGSTONE, I., 'Complementarity and competitiveness of large- and small-scale irrigation farming: A Tanzanian example', Oxford Bulletin of Economics and Statistics, **40,** (3), 1978.

HEALEY, J. M., *The Development of Social Overhead Capital in India, 1950–60* (Blackwell, Oxford, 1965).

HEXEM, R. W. AND HEADY, E. O., *Water Production Functions for Irrigated Agriculture* (Iowa State University Press, Ames, Iowa, 1978).

HOGG, H. G., RANKINE, L. B. AND DAVIDSON, J. R., 'Estimating the productivity of irrigation water', *USDA Agricultural Economics Research*, **22,** (1), January 1970.

HOWE, C. W., *Benefit-cost Analysis for Water System Planning* (American Geophysical Union, Washington, DC, 1971).

IRVIN, G. W., *Irrigation planning at farm level using linear programming techniques*, IDS Discussion Paper 18 (University of Sussex, 1973).

LAL, DEEPAK, *Wells and Welfare*—an exploratory cost-benefit study of the economics of small-scale irrigation in Maharashtra Development Centre of OECO (Paris, 1972).

LITTLE, I. M. D. AND MIRRLEES, J. *Project Appraisal and Planning for Developing Countries* (Heinemann, London, 1974).

MAASS, ARTHUR, MAYNARD, M. HUFSCHMIDT, DORFMAN, ROBERT,

THOMAS, HAROLD A. JR., MARGLIN, STEPHEN A. AND FAIR, GORDON MASKEW, *Design of Water Resource Systems* (Harvard University Press, Cambridge, Mass, 1962).

MAJOR, D. C., *Multi-objective Water Resource Planning,* American Geophysical Union, Water Resources Monograph 4, 1977.

MINHAS, B., PARIKH, K. S. AND MARGLIN, S. A., *Scheduling and Operations of the Bhakra System: Studies in Technical and Economic Evaluation* (Statistical Publishing Company, Calcutta, India, 1972) 89 p.

——, Discussion Paper 59 (Indian Statistical Institute, New Delhi) and 'Toward the structure of a production function for wheat yields with dated inputs of irrigation water', *Water Resources Research*, **10,** (3), 1974.

MINISTRY OF OVERSEAS DEVELOPMENT, 'A guide to the economic appraisal of projects in developing countries' (HMSO, London, 1977).

MITRA, A. K., 'An economic appraisal of farm management and cropping patterns'. Report prepared by FAO for Government of Iraq: Pilot Project in Greater Mussayib Area (FAO, Rome, 1975).

MOHAMMAD, GHULAM, 'Waterlogging and salinity in the Indus Plain: a critical analysis of some of the major conclusions of the Revelle report', *Pakistan Development Review*, **4,** (3), 357–403, 1964 (Karachi).

——, 'Private tube-well development and cropping patterns in West Pakistan', *Pakistan Development Review,* **5,** 1965.

——, 'Development of irrigated agriculture in East Pakistan: some basic considerations', *Pakistan Development Review*, **6,** (3), 315–75, Autumn 1966.

MONDAL, G. C. AND ROY, N. K., 'A cost-benefit analysis of deep and shallow tubewells in West Bengal', *Economic Affairs*, **19,** (3), 1974.

MOORE, C. V., 'A general framework for estimating the production function for crops using irrigation water', *Journal of Farm Economics,* **43,** 876–88, 1961.

—— AND HEDGES, T. R., 'Economics of on-farm irrigation water availability and costs and related farm adjustments', California Agricultural Experiment Station, Giannini Foundation Research Report 263, vol. 2 (University of California, Berkeley, 1963).

MOORTI, T. V. AND MELLOR, J. W., 'A comparative study of costs and benefits of irrigation from State and private tubewells in Uttar Pradesh', *Indian Journal of Agricultural Economics,* **4,** 181–8, 1973.

NORTON, R. D., BASSOCO, L. M. AND SILOS, J. S., *Appraisal of Irrigation Projects and Related Policies and Investments* (Ministry of the Presidency, Mexico, 1974), mimeo WP184, 29 p. (World Bank, Washington).

PALMER-JONES, R., 'Some aspects of the economics of irrigating tea in Malawi', unpublished Ph.D. thesis, University of Reading, 1975.

——, 'Estimating irrigation crop response from data on unirrigated crops', *Journal of Agricultural Economics*, **58,** (1), 85–7, 1976.

——, 'Irrigation system operating policies for mature tea in Malawi', *Water Resources Research*, **13,** (1), 1977.

PIZADUMAN, K., 'An economic study of water management programme in Sambalopur District, Orissa', *Indian Journal of Agricultural Economics*, **24,** (2), 1974.

POULEQUIN, L., 'Risk analysis in project appraisal', Occasional Paper 11 (World Bank, Washington, DC, 1971).

REUTLINGER, S. AND SEAGRAVES, J., 'A method of appraising irrigation returns', *Journal of Farm Economics,* **44,** August 1962.

ROGERS, P. AND SMITH, D. U., 'Integrated use of surface and groundwater in irrigation project planning', *American Journal of Agricultural Economics*, February 1970.

RUTTAN, V. R., 'Induced innovation and agricultural development', *Food Policy*, **2,** (5), 196–216, 1977.

SAGARDOY, J., *Water Charges for Agriculture in Some Selected Countries*, mimeo (FAO, 1973).

SQUIRE, L. AND VAN DER TAK, H. G., *Economic Analysis of Projects* (John Hopkins University, Baltimore, Md, 1975).

STONER, R. F. AND MILNE, D. M., *Economic Design of Wells in Deep Aquifers* (International Commission for Irrigation and Drainage, Moscow, 1974).

STULTS, H. M., 'Predicting farmer response to a falling water-table: an Arizona Case Study', in *Water Resources and Economic Development of the West*, Report 15, Conference proceeding (Committee on the Economics of Water Resources Development of the Western Agricultural Economics Research Council, Las Vegas, December 1966).

TAYLOR, D. C. *Agricultural Development Through Group Action to Improve the Distribution of Water in Asian Gravity Flow Irrigation Systems*, Teaching and Research Forum 1, The Agricultural Development Council Inc. (Singapore University Press, Singapore, 1976).

THORMANN, P. H., 'Labour-intensive irrigation works and dam construction—a review of some research', *International Labour Review*, **106,** (2–3), 151–67, August–September 1972.

THORNER, D., 'The weak and the strong on the Sarda canal' in *Land and Labour in India* (Asia Publishing House, 1962).

THORNTON, D. S., 'Some aspects of the organization of irrigated areas', *Agricultural Administration*, **2,** (3), 79–94, 1975.

THRALL, R. M. (ed.), *Economic Modelling for Water Policy Evaluation* (North Holland Publishing Co., Amsterdam, 1976).

UNIDO, *Guidelines for Project Evaluation* (United Nations, New York, 1972).

UPTON, M. *Irrigation in Botswana*, Reading University, Department of Agricultural Economics Department Study 5, 1969.

WADE, ROBERT, 'Administration and the distribution of irrigation benefits', *Economic Political Weekly*, **x,** (44–5), 1743–7, November 1975.

——, 'Performance of irrigation projects', *Economic Political Weekly,* **xi** (3), 63–6, January 1976a.

——, 'How not to redistribute with growth: the case of India's command area development programme', *Pacific Viewpoint,* **17** (2), 96–104, September 1976b.

WARFORD, J. J. AND JULIUS, S. DEANNE, 'The multiple objectives of water rate policy in less developed countries', *Water Supply and Management*, **1,** (3), 1977 (Pergamon Press, Oxford).

WORLD BANK STUDY GROUP, *Water and power resources of West Pakistan: a study in sector planning*, vol. i, The Main Report; vol. ii, The Development of Irrigation and Agriculture; vol. iii, Background and Methodology (The Johns Hopkins University Press, vol. i, 1968).

YARON, D., *Economic Criteria for Water Resource Development and Allocation* (Hebrew University of Jerusalem, 1966).

——, SHALHEVER, J. AND BRESLER, E., *Economic evaluation of water salinity in irrigation* (The Hebrew University, Faculty of Agriculture, Regovot, Israel, 1974).

BIBLIOGRAPHY AND DICTIONARIES

A bibliography on response is presented in Anderson, Raymond L. and Maass, A., *A simulation of irrigation systems*, Technical Bulletin 1431, ERS, USDA (revised edition 1974).

BERNARD-KIRUKHINE, B., *Water Resources Development and Public Health* (World Health Organization, 1975).

'Bibliography on interregional water transfers', *Water Supply and Management,* **2,** 187–211, 1978 (Pergamon Press, Oxford).

BROWN, R. J. *Water Resources in Arid and Semi-arid Regions: a Bibliography with Abstracts*, NTIS. PS-75/737 (Springfield, Va, USA, 1975).

CASEY, H. H., *Salinity Problems in Arid Lands Irrigation: A Literature Review and Selected Bibliography*, 986 citations (University of Arizona, Tuscon, USA, 1972) 300 p.

COMMONWEALTH BUREAU OF SOILS, *Plant Water Requirements: Annotated Bibliography SB 1749* (Rothamstead Experimental Station, 1976).

COWARD, E. W., JR., *Irrigation Institutions and Organizations: An International Bibliography*, Cornell International Agricultural mimeography 49 (Ithaca, New York, 1976).

DAVIS, L. G., *Irrigation and Water Systems in Africa*, A bibliography (Council of Planning Librarians, Illinois, USA, 1977), 1206 p.

ESPADAS, O. T., *Selected Bibliography on Agricultural Project Evaluation*, incl. appendix, mimeo (World Bank EDI Sem. 1, 1972), 24 p. (Free publication).

INTERNATIONAL INSTITUTE FOR LAND RECLAMATION AND IRRIGATION, *Land and Water Development Selected Literature*, bibliography published quarterly.

Irrigation and Drainage Abstracts (Commonwealth Bureau of Soils, Rothamstead).

Irrigation Efficiency: A Bibliography, Water Resources Scientific Information Centre Bibliography Series, vol. 1, 1973, 418 p.; vol. 2, 1976, 506 p. (Washington, DC, USA).

Irrigation Water, 202 citations—1966 to 1974 (Commonwealth Bureau of Soils, Harpenden, UK, 1974), 33 p.

MEREDITH, D. D., *Bibliography on Optimization of Irrigation Systems*, 1918 citations (University of Illinois, Urbana, USA, 1973), 123 p.

MISKA, J. P., *Irrigation Research: Annotated Bibliography, 1965–74*, 1280 citations (Commonwealth Agricultural Bureau. Farnham Royal, UK, 1976).

NERC, *Hydrological research in the United Kingdom (1970–75)* (Natural Environment Research Council, Alhambra House, 27–33 Charing Cross Road, London, WCH OAX).

OECD, *Water resources selected bibliography*. From an experts' meeting on research on water resources, 17–19 March 1976 (Paris, February 1976).

PICKFORD, J. (ed.), *Indexed Bibliography of Publications on Water and Waste Engineering for Developing Countries* (Loughborough University of Technology, 1977), 50 p.

RADA, L. E. AND BERQVIST, R. J., *Irrigation Efficiencies in Producing Calories and Proteins: An Annotated Bibliography* (University or California, Davis, Ca, USA, 1975), 50 p.

'Register of international rivers', *Water Supply and Management*, **2,** (1), 1978 (Pergamon Press, Oxford).

'Select bibliography on project planning', *IDS Bulletin*, **10,** (1), 1978.

Sewage and Organic Waste Irrigation, bibliography with abstracts (111 abstracts) (National Technical Information Service, Springfield, Va, USA, 1975), 116 p.

SPOONER, B. C., *Irrigation and Employment Strategies* (Asia and Africa), an annotated bibliography (Institute of Development Studies, University of Sussex, 1976).

Sprinkler Irrigation, Especially of Arable Crops and Grassland: Literature, 1965–75 (Centre for Agricultural Publishing and Documentation, Wageningen, The Netherlands, 1976).

Technical Dictionary of Irrigation and Drainage (International Commission on Irrigation and Drainage, Iranian National Committee, Teheran, 1975), 1458 p. (English, French, Persian).

TODD, D. K., *The Water Encyclopaedia* (Water Information Centre, New York, 1970).

VASILIADES, K. C. AND SHANNON, C., *Bibliography for Programming for Agriculture*, 3, mimeo (EDI Sem. 6) (World Bank, Washington, DC, 1973).

Water for Agriculture: Annotated Bibliography, Irrigation and Drainage Paper 22 (Rome, Italy, 1973), 400 p. (English, French, Spanish).

Water Resources Research Centre, Technical Reports Series (University of Hawaii, Honolulu).

World Agricultural Economics and Rural Sociology Abstracts (WAERSA) (Commonwealth Agricultural Bureau, Farnham Royal).

For a review of technical material relating to drip irrigation techniques, see *Proceedings Second International Drip Irrigation Congress* (San Diego, California, July 1974).

JOURNALS AND NEWSLETTERS

Irrigation is a multi-disciplinary activity. Therefore numerous journals contain material related to irrigation development. Among the most specialised journals are:

Water Supply and Management (Pergamon Press, Headington Hill Hall, Oxford OX13 0BW).

Journal of Soil and Water Conservation (Soil Conservation Society of America, 7515 Northeast Street, Ankeny Road, Ankeny, 1A 50021), Max Schnepf (ed.).

Agricultural Water Management (Elsevier Scientific Publishing Co., Box 211, Amsterdam, Netherlands).

Irrigation Science (Springer Verlag, Berlin).

Irrigation Farmer (Gardner Printing & Publishing Pty. Ltd., Box 175, Nunawading, Vic. 3131, Australia), R. A. Gardner (ed.).

Irrigation Journal (Brantwood Publications Inc., P.O. Box 77, Elm Grove, WI 53122), R. W. Morey (ed.).

Water Resources Research (American Geophysical Union, 1909 K Street, NW, Washington, DC2006) R. A. Freeze and D. C. Major (eds).

Water Research (Pergamon Press, Headington Hill Hall, Oxford OX3 0BW).

Irrigation Age (Webb Co., 1999 Shepard Road, St Paul, MN 55116), Ron Ross (ed.).

Various groups circulate newsletters and bulletins on current irrigation developments and research reports. The following are extremely useful sources of information:

Irrinews (Volcani Centre, PO Box 49, Bet Dagan, Israel).

ODI Irrigation Organization and Management Network (Overseas Development Institute, 10/11 Percy Street, London, UK).

From the UNICEF Water Front (UNICEF, New York, N.Y. 10017).

Land, Food and People (Food and Agriculture Organization of the United Nations).

Technical Letter (Land and Water Development Division, FAO, Rome).

F.A.O. International Rice Commission Newsletter (FAO of the United Nations, Agriculture Department, Via delle Terme di Caracalla, 00100 Rome, Italy).

International Rice Research Newsletter (International Rice Research Institute, Manila, Philippines).

Commonwealth Secretariat Rice Bulletin (Commonwealth Secretariat, Marlborough House, Pall Mall, London, SW1Y 5HX, UK).

Asian Regional Irrigation Communication Network (Agricultural Development Council, Box 11 and 172, Bangkok 11, Thailand).

ICID Bulletin (International Commission on Irrigation and Drainage, 48 Nyaya Marg, Chanakyapuri N.D. 110021, India).

Water Resources Bulletin and Water Resources Newsletter (American Water Resources Association, St Anthony Falls Hydraulic Laboratory, Mississippi River at 3rd Avenue, S.E., Minneapolis, MN 55414), W. R. Boggess (ed.).

A full list of agricultural, rural development, irrigation, soil, and other journals is contained in: *Ulrich's International Periodicals Directory*, 17th ed. 1977–8 (Bowker, New York and London, 1977).

AUTHOR INDEX

Bibliographical details of the following authors' works will be found on the pages cited in *italics*.

SUBJECT INDEX